METHODS IN MOLECULAR BIOLOGY

Series Editor
John M. Walker
School of Life and Medical Sciences
University of Hertfordshire
Hatfield, Hertfordshire AL10 9AB, UK

For further volumes:
http://www.springer.com/series/7651

Insect Genomics

Methods and Protocols

Edited by

Susan J. Brown

Division of Biology, Kansas State University, Manhattan, NY, USA

Michael E. Pfrender

Department of Biological Sciences, University of Notre Dame, Notre Dame, IN, USA

Humana Press

Editors
Susan J. Brown
Division of Biology
Kansas State University
Manhattan, NY, USA

Michael E. Pfrender
Department of Biological Sciences
University of Notre Dame
Notre Dame, IN, USA

ISSN 1064-3745 ISSN 1940-6029 (electronic)
Methods in Molecular Biology
ISBN 978-1-4939-8774-0 ISBN 978-1-4939-8775-7 (eBook)
https://doi.org/10.1007/978-1-4939-8775-7

Library of Congress Control Number: 2018958311

This Humana Press imprint is published by the registered company Springer Science+Business Media, LLC, part of Springer Nature.
The registered company address is: 233 Spring Street, New York, NY 10013, U.S.A.

Preface

Insects surround us and play important roles in our ecosystem, from pollinating fruits to vectoring disease. As the founders of the initiative to sequence five thousand genomes (i5K) emphasized, understanding insect genomes will inform insect biology at all levels, from molecular to physiological. When scientists focused on physiology, toxicology, or development decide to sequence the genome of their favorite organism, they must address a raft of questions before they can start, because every step in a genome project presents unique challenges. The content of this volume focuses on methods currently used in sequencing, assembling, and analyzing insect genomes to help graduate students, postdocs, and PIs that are new to genomics navigate this new terrain. (The target audience includes graduate students, postdocs, and research scientists new to the field.).

There are many different ways to approach a genome project, and several strategies are presented in Chapter 1. Genome size must be determined to guesstimate the amount of data required to produce a complete assembly. A protocol for determining genome size by fluorescent measurement in a flow sorter is described in Chapter 2. Although there are specific DNA isolation kits described for most next-generation sequencing technologies that produce ~50 kb DNA, there are circumstances that required much longer starting material, and a HMW DNA extraction protocol in Chapter 3 details the carefully handling of a basic protocol.

Before isolating the DNA, the bioinformatics tools for assembly and analysis should be considered. Understanding the workflow of a genome project and improvements provided by long-range sequencing approaches is discussed in Chapter 4. There are many ways to validate an assembly, and several quality control measures are described in the workflow presented. Beyond genome assembly, a significant challenge for any new genome project is annotating functional elements in the genome. There are a number of ways to annotate genes, and here, we present three chapters to assist with this process. Sometimes we seem to drown in the plethora of data generated as part of the genome project. Picking the best (maybe longest) transcripts from a list of expressed sequences and computer-generated forms is a daunting challenge that exceeds the capacity of any individual to manually annotate all the genes in an insect genome, but implementing specific software tools such as those described in Chapter 5 can produce a helpful set of "best" mRNAs to annotate a genome. An evaluation of different computational approaches to gene annotation is provided in this chapter, and a pipeline for generating automated gene predictions is developed and described. In addition to gene predictions, comparisons with other genomes can help us decide if the genome of interest is close to complete or lacking essential genes. Using an orthology-based framework, the BUSCO tool described in Chapter 6 allows us to assess sets of conserved single-copy genes across clades of insects at different phylogenetic levels. Once an assembly is considered "good enough" and automated gene predictions have been generated, a genome browser, hosted by the i5K workspace at the USDA national agricultural library, can be linked to web tools for manual curation of genes. This critical infrastructure greatly simplifies the organization of the research community most familiar with the biology of the organism under study and facilitates annotation of their favorite genes. Detailed protocols to extract data from this type of database along with quality control procedures which are critical to downstream analysis are presented in Chapter 7.

Access to the full-genome sequence provides a great deal more information than can be inferred from a transcriptome. One goal of many genome projects is to provide a resource that enables a translation of genome structure and content into studies of genome function. Interest in gene regulation begs the identification of promoters and enhancers. Three protocols in Chapters 8–10 describe FAIRE, RAMPAGE, and computational analysis of cis-regulatory modules in insects. Epigenetic modifications have also been implicated in patterns of gene regulation, and insect genomes display a wide range of methylation patterns that affect insect biology in a variety of ways. Chapters 11 and 12 focus on the generation of methylation patterns through whole-genome bisulfite sequencing and computational approaches to analyze these data.

Genome projects can be greatly enhanced by data from independent sources such as physical maps. Genome assemblies can be linked to chromosome-level architecture, and structural arrangements including large inversions can be described through Fluorescent In Situ Hybridization (FISH) techniques. The methods and application of FISH to examine genome structure in mosquito insects are described in Chapter 13.

Novel extensions of NGS sequencing approaches have opened up new avenues to study the microbial communities on and within insects. Many insects carry endosymbionts, the genomes of which are often revealed in the sequence data of the whole insect. In many cases, these microbial partners have profound impacts on the biology of their insect hosts as is the case with *Wolbachia* which can affect reproductive isolation and species divergence. A targeted approach to assay insect genomes (TEEseq) for the presence of endosymbionts is provided in Chapter 14.

The final chapter demonstrates the potential for powerful application of the data derived from insect genomes. In Chapter 15, the sequence of key developmental genes in mosquitos is used to develop an engineered yeast that constitutes a delivery system for lethal RNAi knockdowns. This method, which begins with a complete insect genome sequence, provides a generalizable model for RNAi systems that may find application in many pest and pathogen contexts.

As the price of next-generation sequencing goes down, the power of genomic approaches becomes affordable to virtually every biological inquiry. Along with the increased availability of these data is the need for powerful computational approaches to assemble and annotate insect genomes. The collection of protocols gathered herein is meant to provide an introduction to the workflows and bioinformatics tools available to researchers. These tools and approaches are rapidly evolving, so it is important to check for updates, improvements, and totally new tools that can perform the desired analysis.

We thank Michelle Colemen and Drs. Teresa Shippy and Sherry Miller for assistance in reviewing each chapter.

Manhattan, NY, USA *Susan J. Brown*
Notre Dame, IN, USA *Michael E. Pfrender*

Contents

Contributors

GLEB N. ARTEMOV • *Department of Entomology and Fralin Life Science Institute, Virginia Tech, Blacksburg, VA, USA*
JANA ASSELMAN • *Laboratory of Environmental Toxicology and Aquatic Ecology, Environmental Toxicology Unit (GhEnToxLab), Ghent University, Ghent, Belgium*
ANGELINA BERNARDINI • *Interdisciplinary Program in Genetics, Texas A&M University, College Station, TX, USA*
VOLKER P. BRENDEL • *Department of Biology, Indiana University, Bloomington, IN, USA; School of Informatics and Computing, Indiana University, Bloomington, IN, USA*
SUSAN J. BROWN • *Division of Biology, Kansas State University, Manhattan, KS, USA*
XIAOLONG CAO • *Department of Entomology and Plant Pathology, Oklahoma State University, Stillwater, OK, USA*
MEI-JU MAY CHEN • *Graduate Institute of Biomedical Electronics and Bioinformatics, National Taiwan University, Taipei, Taiwan*
LI-MEI CHIANG • *Graduate Institute of Biomedical Electronics and Bioinformatics, National Taiwan University, Taipei, Taiwan*
CHRISTOPHER P. CHILDERS • *Agricultural Research Service, National Agricultural Library, USDA, Beltsville, MD, USA*
MICHELLE COLEMAN • *Division of Biology, Kansas State University, Manhattan, KS, USA*
MEREDITH M. DOELLMAN • *Department of Biological Sciences, Galvin Life Sciences Building, University of Notre Dame, Notre Dame, IN, USA*
MOLLY DUMAN-SCHEEL • *Department of Medical and Molecular Genetics, Indiana University School of Medicine, South Bend, IN, USA; Eck Institute for Global Health, University of Notre Dame, Notre Dame, IN, USA; Department of Biological Sciences, University of Notre Dame, Notre Dame, IN, USA*
SCOTT P. EGAN • *Department of BioSciences, Rice University, Houston, TX, USA*
JEFFREY L. FEDER • *Department of Biological Sciences, Galvin Life Sciences Building, University of Notre Dame, Notre Dame, IN, USA*
FANNY GATZMANN • *Division of Epigenetics, DKFZ-ZMBH Alliance, German Cancer Research Center, Heidelberg, Germany*
MARC S. HALFON • *Departments of Biochemistry, Biomedical Informatics, and Biological Sciences, University at Buffalo-State University of New York, Buffalo, NY, USA; NY State Center of Excellence in Bioinformatics and Life Sciences, Buffalo, NY, USA; Department of Molecular and Cellular Biology and Program in Cancer Genetics, Roswell Park Comprehensive Cancer Center, Buffalo, NY, USA*
LIMB K. HAPAIRAI • *Department of Medical and Molecular Genetics, Indiana University School of Medicine, South Bend, IN, USA; Eck Institute for Global Health, University of Notre Dame, Notre Dame, IN, USA*
CARL E. HJELMEN • *Department of Entomology, Texas A&M University, College Station, TX, USA*
GLEN R. HOOD • *Department of Biological Sciences, Galvin Life Sciences Building, University of Notre Dame, Notre Dame, IN, USA; Department of Biological Sciences, Wayne State University, Detroit, MI, USA*

HAOBO JIANG • *Department of Entomology and Plant Pathology, Oklahoma State University, Stillwater, OK, USA*
J. SPENCER JOHNSTON • *Department of Entomology, Texas A&M University, College Station, TX, USA*
MAJID KAZEMIAN • *Departments of Biochemistry and Computer Science, Purdue University, West Lafayette, IN, USA*
HAN LIN • *Graduate Institute of Biomedical Electronics and Bioinformatics, National Taiwan University, Taipei, Taiwan*
JACQUELINE A. LOPEZ • *Genomics and Bioinformatics Core Facility, University of Notre Dame, Notre Dame, IN, USA*
FRANK LYKO • *Division of Epigenetics, DKFZ-ZMBH Alliance, German Cancer Research Center, Heidelberg, Germany*
DANIEL J. MCKAY • *Department of Biology, The University of North Carolina at Chapel Hill, Chapel Hill, NC, USA; Department of Genetics, Integrative Program for Biological and Genome Sciences, The University of North Carolina at Chapel Hill, Chapel Hill, NC, USA*
KESHAVA MYSORE • *Department of Medical and Molecular Genetics, Indiana University School of Medicine, South Bend, IN, USA; Eck Institute for Global Health, University of Notre Dame, Notre Dame, IN, USA*
MONICA F. POELCHAU • *Agricultural Research Service, National Agricultural Library, USDA, Beltsville, MD, USA*
R. TAYLOR RABORN • *Department of Biology, Indiana University, Bloomington, IN, USA*
JACOB S. REALEY • *Department of Medical and Molecular Genetics, Indiana University School of Medicine, South Bend, IN, USA; Eck Institute for Global Health, University of Notre Dame, Notre Dame, IN, USA*
STEPHEN RICHARDS • *Human Genome Sequencing Center, Baylor College of Medicine, Houston, TX, USA*
SURYA SAHA • *Sol Genomics Network, Boyce Thompson Institute, Ithaca, NY, USA*
NICHOLAS D. SCHEEL • *Eck Institute for Global Health, University of Notre Dame, Notre Dame, IN, USA; Department of Biological Sciences, University of Notre Dame, Notre Dame, IN, USA*
HANNES SCHULER • *Department of Biological Sciences, Galvin Life Sciences Building, University of Notre Dame, Notre Dame, IN, USA; Faculty of Science and Technology, Free University of Bozen-Bolzano, Bolzano, Italy*
MATHIEU SEPPEY • *University of Geneva and Swiss Institute of Bioinformatics, Geneva, Switzerland*
DAVID W. SEVERSON • *Department of Medical and Molecular Genetics, Indiana University School of Medicine, South Bend, IN, USA; Eck Institute for Global Health, University of Notre Dame, Notre Dame, IN, USA; Department of Biological Sciences, University of Notre Dame, Notre Dame, IN, USA*
IGOR V. SHARAKHOV • *Department of Entomology and Fralin Life Science Institute, Virginia Tech, Blacksburg, VA, USA*
MARIA V. SHARAKHOVA • *Department of Entomology and Fralin Life Science Institute, Virginia Tech, Blacksburg, VA, USA*
FELIPE A. SIMÃO • *University of Geneva and Swiss Institute of Bioinformatics, Geneva, Switzerland*
VLADIMIR A. TIMOSHEVSKIY • *Department of Entomology and Fralin Life Science Institute, Virginia Tech, Blacksburg, VA, USA*

Robert M. Waterhouse • *Department of Ecology and Evolution, University of Lausanne, and Swiss Institute of Bioinformatics, Lausanne, Switzerland*
Na Wei • *Eck Institute for Global Health, University of Notre Dame, Notre Dame, IN, USA; Department of Civil and Environmental Engineering and Earth Sciences, University of Notre Dame, Notre Dame, IN, USA*
Evgeny M. Zdobnov • *University of Geneva and Swiss Institute of Bioinformatics, Geneva, Switzerland*

Chapter 1

Arthropod Genome Sequencing and Assembly Strategies

Stephen Richards

Abstract

As in any endeavor, the strategy applied to a genome project can mean the difference between success and failure. This is especially important when limited funding often means only a single approach may be tried at a given time. Although the advance of all areas of genomics and transcriptomics in recent years has led to an embarrassment of riches, methods in the field have not quite reached the turn-key production status for all species, despite being closer than ever. Here I contrast and compare the technical approaches to genome projects in the hope of enabling strategy choices with higher probabilities of success. Finally, I review the new technologies that are not yet widely distributed which are revolutionizing the future of genomics.

Key words Genome project strategy, Genome assembly, Genome sequencing, Genomics, Insect genomes, Oxford nanopore, Pacific biosciences, 10× Genomics

1 Introduction

Ideally, any biologist needing genomic sequence information would rapidly find it in a public database [1] as is the case for the model organisms and many species with large communities today. Alternatively, a rapid, routine, robust, and low-cost procedure leading to high-quality reference sequence and annotations would rapidly fill in any species gap. This is today the case for prokaryotic genomes and users of model species or key organisms of interest such as the human genome. In this world, researchers utilize genomic infrastructure addressing biological research questions such as connecting change in genotype and phenotype, assessing environmental effects on populations and their genetics, and better understanding the conserved and variable biology of the animal cell. Unfortunately, while this world exists today for some researchers investigating popular species, we have far to go for the remainder of the greater than 1.5 million described species on planet Earth. As sequencing costs have decreased, researchers focusing on non-model organisms are jumping into genome projects themselves, or assigning them to a graduate student (often

Susan J. Brown and Michael E. Pfrender (eds.), *Insect Genomics: Methods and Protocols*, Methods in Molecular Biology, vol. 1858, https://doi.org/10.1007/978-1-4939-8775-7_1,

successfully—*see* [2]). In this article, I outline the strategic choices needed for a genome project and address their historical and current pros and cons. In particular, I place an emphasis on the quest for the longest contigs, and chromosome length scaffolds to underlie future biological research.

As of Oct 20, 2017, 297 arthropod species had some kind of genome assembly in NCBI, representing 33 orders, and only 124 of 2999 arthropod families (http://i5k.github.io/arthropod_genomes_at_ncbi). To fully describe life on earth, there is a lot of work left to do.

2 Major Steps in a Genome Project

There are four major steps in a genome project: (1) Strategy Choice. (2) Appropriate DNA isolation and DNA sequencing to feed the decided genome assembly choice. (3) Genome assembly, annotation, and submission to public databases. (4) Analysis and possible publication.

Strategy Choice depends greatly on the species being studied, and partly on the desired result. Certain taxa have been proven over time to be relatively easy to sequence, assemble, and annotate. The species-specific criteria include genome size, which affects final cost, genome polymorphism, which affects the probability of high-quality assembly or requires methods that address this, and individual size, which determines the minimum number of individuals and thus different haplotypes required for DNA isolation and sequencing. For example, since prokaryotes can be cultured to generate enough DNA, long read sequencing and genome assembly into a "circle" has become the community standard. Prokaryotes have small, single copy genomes, and zero polymorphism, which enables high-quality routine genome assembly. For mammals, a high-quality reference based on 70× PB long reads is becoming the norm, depending on available funds and genome size, but also cheaper "draft" assemblies based on Illumina short reads are effective for many research questions. Although mammals generally have 2.5–3.5 Gb genomes, they have relatively low polymorphism (1 SNP per 1000 bp in humans—below the sequencing error rate) and sufficient DNA can be isolated from a single individual or a single blood draw. Most genome assembly software has been either explicitly designed for mammalian assembly or at least tested on human genomes, and so is tuned for mammalian genomes. Other vertebrate classes seem to follow certain patterns. Bird genomes are generally ~1–1.5 Gb in size, and relatively easy to sequence and assemble. Fish also have small genomes, although sequence polymorphism may be more of an issue. Little is known about reptilian genomes, and amphibian genomes tend to be large, so few have been attempted because of cost. Outside of the vertebrates, things get more problematic. Echinoderms, such as sea urchins and other sea creatures, tolerate high polymorphism—2% SNPs for the sea urchin. When combined with repetitive regions, this makes short read genome sequencing and assembly approaches problematic, generating short contigs, and poor annotation due to assembly gaps. Here, we focus on insects where small physical size in large populations can lead to problems. The small physical size can mean that multiple individuals are required to isolate a sufficient amount of DNA—which can be significant for long read sequencing technologies, and/or long insert libraries, where the first step in the process is shearing or fragmentation of the DNA and size selection of long molecules. Relatively large populations possible in insects enable high genome sequence polymorphism, making traditional genome assembly more likely to produce low contiguity (i.e., contig N50s) and more gaps in assemblies leading to poor gene model annotation.

DNA isolation and sequencing must be appropriate to the strategy choice. Once a strategy is decided on, the DNA must be isolated, and sequence generated appropriately in type and amount to feed to the chosen assembly tool. While this is clearly common sense, all too many labs have generated some Illumina data of their favorite species, given it to the lab person most familiar with a computer, and complained when the desired results were not obtained. Clearly no researcher would do the same within their field of expertise. Some examples can provide an idea of the differences in data feedstocks for different assemblies. First, in a 10× genomics assembly strategy, one ideally isolates high molecular weight DNA (molecules >100 kb in length), from a single individual, uses the 10× genomics DNA partition library construction microfluidics machine, generates 55–60× fold genome coverage with 150 bp paired Illumina reads, and assembles using the company's Supernova assembly software [3]. Second, in a PB long read genome assembly strategy, one also isolates relatively long DNA molecules, then shears it and isolates fragments in the 20–30 kb range to construct libraries, which are sequenced to 50–70× genome coverage using a PB single molecule sequencing machine. Assembly is then performed using publically available software such as Canu [4] by first error correcting the longest reads using shorter PB reads, and then using 15–25× error corrected long read data for the final assembly. Finally, the standard Illumina De Bruijn genome assembly strategy requires multiple sized insert Illumina paired end libraries (e.g., 500 bp, 2 kb, 5 kb, and 8–10 kb insert sizes) each sequenced to 20–40× genome coverage, fed to a De Bruijn short read genome assembler such as soap de-novo [5] or allpaths LG [6]. The final genome coverage might be 150×, and results will be somewhat dependent on the polymorphism and repeat content of the species genome.

Bacterial contaminants. A final question is whether or not to isolate DNA with microbial symbionts present or to try to remove them by dissection, or antibiotic treatment. Experience from the i5K pilot project (30 arthropod species sequenced using an Illumina approach), has found it best to keep bacterial symbionts be they obligate in special cells, or simply living in the insect gut when performing DNA isolation. Often the bacterial DNA assembles quite well (since it lacks polymorphism or significant repeats) and is trivial to computationally identify and separate in the final assembly. Thus it is best to simply keep interesting symbiont information. Some amount of starvation prior to DNA isolation to allow the digestion of gut contents can be a good idea,—especially in cases such as ticks where body size can significantly change during feeding—but usually the amount of contaminating DNA sequence is not noticed at all. Note that it is critical to understand the context of the bacteria. In the bedbug genome [7] project, they sequenced a laboratory culture fed on rabbit blood through a laboratory filter. However, to encourage feeding, the filter was rubbed on human skin, which increases feeding success in this system. A human skin bacterial commensal genome with sequence similarity to *S. xylosus* was found in the resulting sequence, presumably living in the gut of the lab culture, but the biological significance is hard to assess.

Genome assembly. The primary software dictated by the sequencing and assembly strategy can simply be run with default parameter to generate an initial assembly, but can be run multiple times to identify the incrementally best assembly result, greatly increasing assembly compute times. Many genome assemblers require large RAM machines (500 Gb—1 Tbyte RAM for De Bruijn graph assemblers) depending on genome size. Access to this hardware was a significant impediment for many labs, but prices of large RAM equipment have decreased dramatically, and large RAM machines can be rented on Amazon Web Services (AWS) and other commercial clouds. Mammalian genome assemblies using the cloud can easily cost > $500 at the current time, assuming no significant difficulties and a single straightforward execution. Rerunning the assembly multiple times with different options or different inputs increases both in financial and time costs. Assembly times have dramatically decreased over time. Mammalian genome assembly using Allpaths LG on a 1 Tb RAM machine with 32 cores took ~2–3 weeks in 2014, today in 2017, the same mammalian genome assembly using the 10× genomics supernova assembler requires less memory (<500 Gb) and, using similar hardware, completes in 2–3 days [3]. As soon as multiple assemblies obtain similar QC metrics (for example, CEGMA or BUSCO gene statistics) the urge for incremental improvement should be resisted in favor of biological analysis unless significant new assembly data is added.

It doesn't count unless it is in a public database. Raw species sequence data needs to be submitted to a public repository such as the NCBI Short Read Archive (SRA). Genome assemblies need to be submitted as soon as possible to the NCBI or similar International Nucleotide Sequence Database Collaboration (INSDC). The potential to be scooped on publication of a new genome is low today, instead collaboration and data sharing help to understand the mass of data that is a genome and prevent duplication of work. The Ft. Lauderdale rules [8] designed for the human genome still inspire but are often forgotten, as not all researchers, and importantly journal editors are aware of them. The basic principles of rapid pre-publication data release, and letting data generators publish their own data first are as sound as ever.

Automated annotation has not significantly advanced over the last decade. However, the generation of transcript data critical for automated annotation is easier than ever. RNAseq costs have decreased, and 5 Gb of RNAseq data from both sexes of a species can give significant transcript data for >95% of genes. Common pipelines include Maker [9], NCBI's Refseq, and the ensemble genome annotator, but all are hard to install, and the NCBI and ensemble pipelines are best run by the respective institutions—perhaps at no cost to the researcher in the case of genomes submitted to the NCBI. Instead, it might be best to simply run a simple gene predictor such as AUGUSTUS [10] that integrates RNAseq data, or to use a more modern simplified genome annotation pipeline like "Just Annotate My genome" (http://jamg.sourceforge.net) avoiding the complexity of traditional pipelines. Quality control using a set of known sequences such as BUSCO [11] or CEGMA [12] is key to measure progress. The future, however, looks exciting as long read technologies can now sequence full-length RNA transcripts, and direct RNA sequencing is now possible on the ONT platform. These long read technologies provide high-quality evidence for curation of multiple transcripts per gene with the correct splicing for each of the transcripts. Gene modeling software will likely need to be modified to best take advantage of these improved datasets.

Research group manual curation. Poelchau describes the i5K workspace [13] in this volume, which includes the WebApollo [14] browser critical for the viewing and communication of genome sequence annotations that are not easily accessible in other formats. The critical aim of human curation is to add other information that the automated pipelines do not use—for example, knowledge of gene and protein structure beyond simple RNA and protein sequence alignments—and to ensure published biological statements about genes and gene families are correct. The most common automated annotation errors are the erroneous merging or splitting of gene models. Example statements that need to be checked manually before publication include statements about the absence of a gene, where gene models, full assemblies, an un-assembled reads all need to be searched for the gene in question; the number of genes in a family of interest needs to be confirmed, and can be particularly difficult in gene families with high sequence diversity such as olfactory receptors; and that genes of interest have full gene models rather than partial or split gene models—often caused by a scaffold gap in the assembly where the unit of automated annotation is the sequence scaffold.

Analysis and publication. Genomes of previously un-sequenced species are no longer interesting enough for high impact publication. Instead, the focus of journals has rightly returned to biological analysis and better understanding of the life style and phenotype of the species of interest—usually requiring additional experimental work in addition to the genome sequencing. Functional work has been revolutionized by CRISPR-Cas9, and the cost of generating population genetics or comparative RNA sequence data for alignment to a reference is low and continuing to decrease. Overall, the experimental tools to utilize a genome to address a biological question are better than ever. Thus researchers should not obsess over genome assembly perfection unless the data allows significant improvement. Time spent on incremental genome improvement freezes and new annotations is time not investigating biological questions.

A quick history of sequence technologies and genome assembly strategies—The quest for the chromosome contiguity.

Initial Sanger assemblies focused on highly selected species due to extreme cost—and often had lower contig N50s as costs would demand assembly of lower fold genome coverage. 6× rather than 12× coverage of a mammalian genome saved millions of dollars. Few genomes were "finished" (again for cost reasons), so draft sequences with 10–30 kb contig N50s were and are common. End sequencing of BAC (bacterial artificial chromosome) clones in addition to other insert sizes (a usual strategy included 2–3 kb, 8 kb, fosmid sized inserts, and BAC end sequencing) generally produced scaffold N50s of >1 Mb—which could be extended by further work such as genetic and physical mapping.

454 sequence-based assemblies were similar to Sanger and used similar assembly software approaches [15], but they were much cheaper, enabling appropriate coverage to be generated. However, contigs were not significantly longer as higher coverage was offset by shorter 454 pyrosequencing reads (450 bp vs. 650 bp Sanger read lengths) and expanded application of the technique to more problematic non-model species with less or no inbreeding.

Initial Illumina assemblies required De Bruijn graph assemblers [5, 16, 17] and higher overall fold genome coverage (100× +) due to the short read lengths, but significantly lower costs enabled far more species to be sequenced. Inbred model species generated reasonable assemblies, but the assemblers still perform poorly on highly polymorphic species. The assemblathon contest [18] highlights fairly mediocre results in a relatively easy set of vertebrate non-model species. Results are often worse in outbred insects. Still these techniques have generated good scientific data for many species. Some groups have developed tools to improve assembly of polymorphic sequence data. Platanus [19] and Redundans [20] are two polymorphic genome assembly tools with varying levels of assembly success. This author has generated significant improvements of polymorphic genome contiguity using Redundans.

More recently, the genome assembly field has split into two approaches, each with benefits and costs. The highest quality approach is long read assembly based on PB or ONT data that can generate extremely long contigs with N50s > 10 Mb. Full circular chromosomes are now the standard in bacterial genome sequencing. However, the constraint to date has been cost of reads, further compounded by the cost penalty of higher coverage needed to overcome low read sequence accuracy. The alternative approach is to improve the use of cost-effective Illumina data with innovative sequencing library construction techniques. These include genome partitioning (Complete Genomics Inc), high coverage subclone barcoding and subassembly followed by full genome assembly (Moleculo [21] now Illumina synthetic long reads [22]), and low coverage barcoded genome partitioning (10× genomics [3]). *See* Table 1 for detail on these approaches.

Finally, some new technologies generate different datatypes addressing chromosome level scaffolding and/or genome validation. These include optical mapping [23], BioNano [24], HiC assembly [25], and "Chicago library" variations on HiC offered by Dovetail genomics [26]. Together these technologies all promise the possibility of genome assembly validation and chromosome level super scaffolding of sequences, but all have error rates that are difficult to detect without other data types at the time of writing. In many cases the tools require handholding and experimentation for the best results.

3 Current Genome Sequencing and Assembly Practices

Table 1 summarizes the current choices in genome sequencing and assembly. At this time the major impediments are access to equipment such as the 10× genomics machine or Illumina sequencers through core labs and services, or in the case of the long read strategies, cost and DNA availability and quality. The final experimental design may be determined more by the physical size of the individuals of the species (midges vs. mantis) affecting the quantity of DNA available from a single individual, or genome size, which affects cost or, if all else fails, a kitchen sink approach in the quest for the highest quality result (Fig. 1).

Table 1
Current genome sequencing and assembly practices

Strategy	Assembler	Does it work?	Pros	Cons
70× Pacific Biosciences Long read sequencing	Canu [4] HGAP [27] wtdgb (https://github.com/ruanjue/wtdbg-1.2.8) Falcon [28]	Generally! Proven for prokaryotes, insects, mammals, not sure about the largest genomes or most polymorphic datasets	Long reads → Long contigs—generally high-quality genome assemblies result	Needs a lot of high-quality DNA, Cost (~$200/Gb raw seq on a PB Sequel platform Oct 2017) Final seq needs polishing—but Illumina is relatively cheap for this purpose Need access to a PB Sequel
Oxford Nanopore Long reads	All the PB long read assemblers	We think so… Generally thought to have similar success to PB—has not been tested on so many species as newer	Low capital cost Seq "In lab"—no need for a core lab Possibility for **very** long reads at the expense of yield which could transform the field Higher 1D read accuracy than PB (~92% Oct 2017 vs. 87% for PB)	As of Oct 2017, still an immature technology Needs even more DNA than PB Yield more dependent on DNA quality than PB Best Yields are with 7–10 kb average read lengths—but best assemblies depend on higher numbers of 20 kb + reads
10× genomics library and Illumina reads	Supernova [3]	Generally—some small technical problems with smaller genomes which will be addressed	Low cost, < $3000 for a mammalian sized genome Requires small amounts of DNA—useful for small bodied insects Partitioned assemblies are more reliable than previous Illumina approaches Supernova assembler requires less RAM than previous De Bruijn assemblers	Smaller contigs relative to long read approaches—100 kb in human assemblies HMW DNA required (ideally >100 kb, 50 kb fragments usable) Need access to a 10× genomics machine and the required Illumina HiSeqX platform for 150 bp pe data at lowest cost
HiC Illumina assembly	Juicebox [25]	Publications suggest scaffolding can be chromosome length, but	Possible chromosome level scaffolds albeit with some error	Scaffolding only, add on to standard Illumina approaches

strategies Dovetail Genomics (Santa Cruz CA)	LACHESIS [29]	contig N50s no better the underlying assembly	Relative low cost due to Illumina approach	HiC Library construction not trivial, and requires tissue or cells to make Scaffolding errors exist and are hard to identify without other data
Traditional Illumina approaches Can fill gaps with low coverage long read approaches	Soap de-novo [5] allpathsLG [6] Discovar [30] PB Jelly long read gap fill [31] For dealing with heterozygosity Platanus [19] Redundans [20]	Not well suited for polymorphic outbred species (but try platanus and redundans)—great for mammals, birds, fish, male haploid species such as ants and bees	Lower cost—but not the cheapest (due to multiple libraries and higher required coverage) When it does work—works fine, and is reliable—for example just fine for mammals and birds, and other major taxa where already proven	Multiple required libraries with different insert sizes add to cost, and need for 100× + coverage Doesn't work with polymorphic genomes, leading to some of the above approaches May need software tools such as Redundans to improve poor assemblies

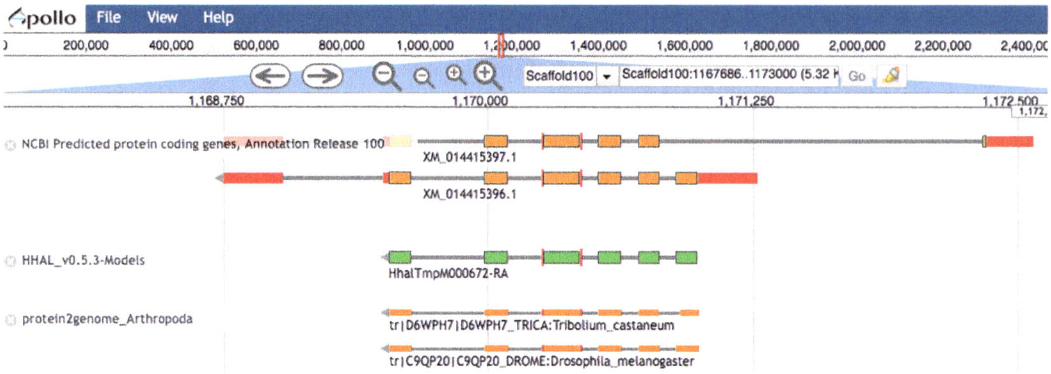

Fig. 1 An automatically generated gene model in the Brown Marmarated Stinkbug Web Apollo Genome Browser at the US National Agricultural Library's Web Apollo Browser. Top track: Orange (coding) and Red (UTR) Gene Model is the NCBI RefSeq predicted protein coding gene. Middle track: Green: Maker 2 Models generated—note much less UTR detail and a target of one gene model per gene. Bottom track: Protein alignment evidence for the viewer to assess biological evidence prior to manual curation of a gene model, which might involve joining or splitting of artificially split or joined gene models, or simply naming a gene model

4 Future of Insect Genomics

Science fiction writer, William Gibson has been quoted as saying "The future is here, but just not evenly distributed"—appropriate for the current state of insect genomics. New technologies promising to reduce the cost of genome sequencing, increase the quality and ease of genome assembly, and extend genome assemblies from telomere to centromere to telomere are just now being released by companies.

10× genomics assemblies have 100+ kb contig N50s and ~10 Mb Scaffold N50s on mammals—at low cost for a high-quality draft [3]. The technique requires 50–60× Illumina sequence coverage of the single 10× library (reagent costs ~\$200—Nov 2017). Thus a draft 3 Gb mammalian genome assembly could cost <\$3000. For smaller genomes there are some issues with the partitioning dilutions which will likely be quickly fixed, but the big advantages of requiring only 1.5 ng DNA (although QC takes more) and the separation of haplotypes in the assembly makes the technique highly attractive for sequencing large numbers of physically small species.

Oxford Nanopore Technology (ONT) sequencing is now available, but a little frustrating given the promise. However, while yield is still a cost issue, the read lengths can be very long. The longest so-called "Whale" read reported aligned human read to date is 882 kb (Loman lab http://lab.loman.net/2017/03/09/ultrareads-for-nanopore/ and [32]) using traditional DNA isolation methods. At the time of writing, researchers are good naturedly racing for the first 1 Mb sequence read, and the first full chromosome length reads. Unfortunately, these protocols generate lower yield and are thus more expensive relative to protocols targeting modest read lengths. Future flow cells with larger pore numbers will decrease the cost. The original long read single molecule sequencing company Pacific Biosciences is still strongly competing especially on price and reliability, but with less outward fanfare, yet a longer history of on time product delivery.

Finally, looking further into the future, it is possible that sequence costs may drop another order of magnitude. Multiple companies are now publically discussing the $100 human genome. Note that presently, significant bulk reagent buys are required to get the lowest sequencing costs, but sequencing company competition is increasing.

Illumina's NovaSeq is arriving in genome sequencing centers and cores, and the company claims the machine has the headroom to reach to $100 human genome. Illumina is likely to be the low cost sequence provider in the short term. At the time of writing (fall 2017) the S4 flowcell was just becoming available and will likely reduce the cost of a human genome to ~$800 (likely $1200 when other costs are accounted for) from the current $1000 (~$1500 accounting for other costs), the S6 flowcell for further capacity and price decrease is not yet available.

In direct competition with Illumina is BGIseq, the BGI's sequencing platform based on their acquisition of Complete Genomics. The rolling circle "nanoball" DNA array technology is fully capable of extreme template densities, and 250 bp paired end reads are also being discussed. Because of business strategy differences BGI may beat Illumina to the $100 human genome.

In the medium term (5 years), ONT is promising to produce larger and larger nanopore arrays. The current MinION flowcells have 512 pore arrays, and 3000 pore PromethION flowcells are just now available in Fall 2017, reducing costs of ONT long read sequencing by ~ fourfold. Tantalizingly, during a company video presentation [33] CIO Clive Brown discussed research on a one million pore chip—which would enable extremely high throughput sequencing, but is currently multiple years from commercialization.

5 Summary

The ultimate aim for biologists must be to have reference sequences for extant species in public databases. Too many investigators are hamstrung in their experiments by the lack of a genome reference. New tools have reduced, and continue to reduce the price of a genome reference to be accessible for many more labs. However, these genome references *must be placed in the public databases*, the ex-students lab-book and outdated computer are no place to keep record of life on planet earth.

Community workspaces with long-term funding are vital. For *Insecta*, the i5K workspace is a vital community space with a longer term view in the National Agriculture Library. There is a long-term need for a fully featured genome browser and community workspace for all species on earth. The major public databases such as NCBI should in theory already be the perfect places for such display of genome reference sequences, but NCBI has long had a poor browser for viewing and no tools for community annotation.

Recommended strategies for insects. (1) Choose a long read strategy with a focus on inputting the longest sequence reads into the assembler if money and DNA amount are not limiting. Between ONT and PB at the current time the ONT technology can be placed in a regular lab for low capital cost and has higher accuracy, but requires significant DNA and is less consistent generating yield. (2) Choose a short read strategy such as 10× genomics for low cost draft genomes generated from small amounts of DNA. Access to the machine is currently expanding and the sequencing will also likely require a core lab or commercial service. (3) If additional DNA and funds are available (but not enough for a full long read strategy) long read gap filling (such as PBJelly) or a HiC or dovetail genomics assembly improvement can be considered. These recommendations will change rapidly with the ongoing evolution of sequencing technologies.

Acknowledgments

I thank Susan Brown and Michael Pfrender for the invitation to author this article. This work was funded by NHGRI grant U54 HG003273 to Richard A. Gibbs.

References

1. Pennisi E (2017) Biologists propose to sequence the DNA of all life on Earth. Science. https://doi.org/10.1126/science.aal0824
2. Kaiser TS, Poehn B, Szkiba D, Preussner M, Sedlazeck FJ, Zrim A et al (2016) The genomic basis of circadian and circalunar timing adaptations in a midge. Nature 540(7631):69–73. https://doi.org/10.1038/nature20151
3. Weisenfeld NI, Kumar V, Shah P, Church DM, Jaffe DB (2017) Direct determination of diploid genome sequences. Genome Res 27(5):757–767. https://doi.org/10.1101/gr.214874.116
4. Koren S, Walenz BP, Berlin K, Miller JR, Bergman NH, Phillippy AM (2017) Canu: scalable and accurate long-read assembly via adaptive k-mer weighting and repeat separation. Genome Res 27(5):722–736. https://doi.org/10.1101/gr.215087.116
5. Li R, Li Y, Kristiansen K, Wang J (2008) SOAP: short oligonucleotide alignment program. Bioinformatics 24(5):713–714. https://doi.org/10.1093/bioinformatics/btn025
6. Maccallum I, Przybylski D, Gnerre S, Burton J, Shlyakhter I, Gnirke A et al (2009) ALLPATHS 2: small genomes assembled accurately and with high continuity from short paired reads. Genome Biol 10(10):R103. https://doi.org/10.1186/gb-2009-10-10-r103

7. Benoit JB, Adelman ZN, Reinhardt K, Dolan A, Poelchau M, Jennings EC et al (2016) Unique features of a global human ectoparasite identified through sequencing of the bed bug genome. Nat Commun 7:10165. https://doi.org/10.1038/ncomms10165
8. Welcome_Trust (2003) Sharing data from large-scale biological research projects: a system of tripartite responsibility Washington DC: The Welcome Trust. https://www.genome.gov/pages/research/wellcomereport0303.pdf
9. Holt C, Yandell M (2011) MAKER2: an annotation pipeline and genome-database management tool for second-generation genome projects. BMC Bioinformatics 12:491. https://doi.org/10.1186/1471-2105-12-491
10. Keller O, Kollmar M, Stanke M, Waack S (2011) A novel hybrid gene prediction method employing protein multiple sequence alignments. Bioinformatics 27(6):757–763. https://doi.org/10.1093/bioinformatics/btr010
11. Simao FA, Waterhouse RM, Ioannidis P, Kriventseva EV, Zdobnov EM (2015) BUSCO: assessing genome assembly and annotation completeness with single-copy orthologs. Bioinformatics 31(19):3210–3212. https://doi.org/10.1093/bioinformatics/btv351
12. Parra G, Bradnam K, Korf I (2007) CEGMA: a pipeline to accurately annotate core genes in eukaryotic genomes. Bioinformatics 23 (9):1061–1067. https://doi.org/10.1093/bioinformatics/btm071
13. Poelchau M, Childers C, Moore G, Tsavatapalli V, Evans J, Lee CY et al (2015) The i5k Workspace@NAL—enabling genomic data access, visualization and curation of arthropod genomes. Nucleic Acids Res 43 (Database issue):D714–D719. https://doi.org/10.1093/nar/gku983
14. Lee E, Helt GA, Reese JT, Munoz-Torres MC, Childers CP, Buels RM et al (2013) Web Apollo: a web-based genomic annotation editing platform. Genome Biol 14(8):R93. https://doi.org/10.1186/gb-2013-14-8-r93
15. NewblerKJ (2005) 454 Life Sciences. Sequence read assembly software. p 454
16. Zerbino DR, Birney E (2008) Velvet: algorithms for de novo short read assembly using de Bruijn graphs. Genome Res 18 (5):821–829. https://doi.org/10.1101/gr.074492.107
17. Butler J, MacCallum I, Kleber M, Shlyakhter IA, Belmonte MK, Lander ES et al (2008) ALLPATHS: de novo assembly of whole-genome shotgun microreads. Genome Res 18 (5):810–820. https://doi.org/10.1101/gr.7337908
18. Bradnam KR, Fass JN, Alexandrov A, Baranay P, Bechner M, Birol I et al (2013) Assemblathon 2: evaluating de novo methods of genome assembly in three vertebrate species. GigaScience 2(1):10. https://doi.org/10.1186/2047-217X-2-10
19. Kajitani R, Toshimoto K, Noguchi H, Toyoda A, Ogura Y, Okuno M et al (2014) Efficient de novo assembly of highly heterozygous genomes from whole-genome shotgun short reads. Genome Res 24(8):1384–1395. https://doi.org/10.1101/gr.170720.113
20. Pryszcz LP, Gabaldon T (2016) Redundans: an assembly pipeline for highly heterozygous genomes. Nucleic Acids Res 44(12):e113. https://doi.org/10.1093/nar/gkw294
21. Voskoboynik A, Neff NF, Sahoo D, Newman AM, Pushkarev D, Koh W et al (2013) The genome sequence of the colonial chordate, *Botryllus schlosseri*. elife 2:e00569. https://doi.org/10.7554/eLife.00569.
22. McCoy RC, Taylor RW, Blauwkamp TA, Kelley JL, Kertesz M, Pushkarev D et al (2014) Illumina TruSeq synthetic long-reads empower de novo assembly and resolve complex, highly-repetitive transposable elements. PLoS One 9 (9):e106689. https://doi.org/10.1371/journal.pone.0106689
23. Jing J, Reed J, Huang J, Hu X, Clarke V, Edington J et al (1998) Automated high resolution optical mapping using arrayed, fluid-fixed DNA molecules. Proc Natl Acad Sci U S A 95(14):8046–8051
24. Shelton JM, Coleman MC, Herndon N, Lu N, Lam ET, Anantharaman T et al (2015) Tools and pipelines for BioNano data: molecule assembly pipeline and FASTA super scaffolding tool. BMC Genomics 16:734. https://doi.org/10.1186/s12864-015-1911-8
25. Dudchenko O, Batra SS, Omer AD, Nyquist SK, Hoeger M, Durand NC et al (2017) De novo assembly of the Aedes aegypti genome using Hi-C yields chromosome-length scaffolds. Science 356(6333):92–95. https://doi.org/10.1126/science.aal3327
26. Putnam NH, O'Connell BL, Stites JC, Rice BJ, Blanchette M, Calef R et al (2016) Chromosome-scale shotgun assembly using an in vitro method for long-range linkage. Genome Res 26(3):342–350. https://doi.org/10.1101/gr.193474.115
27. Chin CS, Alexander DH, Marks P, Klammer AA, Drake J, Heiner C et al (2013) Nonhybrid, finished microbial genome assemblies from

long-read SMRT sequencing data. Nat Methods 10(6):563–569. https://doi.org/10.1038/nmeth.2474
28. Chin J (2013) FALCON
29. Burton JN, Adey A, Patwardhan RP, Qiu R, Kitzman JO, Shendure J (2013) Chromosome-scale scaffolding of de novo genome assemblies based on chromatin interactions. Nat Biotechnol 31(12):1119–1125. https://doi.org/10.1038/nbt.2727
30. Jaffe DB (2015) DISCOVAR: Assemble genomes, find variants. Broad Institute: Broad Institute. https://www.broadinstitute.org/software/discovar/blog. Accessed 25 May 2016
31. English AC, Richards S, Han Y, Wang M, Vee V, Qu J et al (2012) Mind the gap: upgrading genomes with Pacific Biosciences RS long-read sequencing technology. PLoS One 7(11): e47768. https://doi.org/10.1371/journal.pone.0047768
32. Jain M, Koren S, Quick J, Rand AC, Sasani TA, Tyson JR et al (2018) Nanopore sequencing and assembly of a human genome with ultra-long reads. Nat Biotechnol 36:338–345
33. Oxford_Nanopore_Technologies (2016) A technology update from Clive Brown, September 2016—Wafer thin update. Oxford UK

Chapter 2

Genome Size Estimation and Quantitative Cytogenetics in Insects

J. Spencer Johnston, Angelina Bernardini, and Carl E. Hjelmen

Abstract

With care, it is possible using flow cytometry to create a precise and accurate estimate of the genome size of an insect that is useful for genomics, genetics, molecular/cell biology, or systematics. Genome size estimation is a useful first step in a complete genome sequencing project. The number of sequencing reads required to produce a given level of coverage depends directly upon the 1C amount of DNA per cell, while an even more critical need is an accurate 1C genome size estimate to compare against the final assembly. Here we present a detailed protocol to estimate genome size using flow cytometry. Published genome size estimates should be submitted to genomesize.com so that they are available to all.

Key words Flow cytometry, Chromatin structure, Endopolyploidy, Underreplication, Chromosomal sex determination, Sexual dimorphism, Ploidy, Cryptic species

1 Introduction

Accuracy and precision are important in genome size estimates. Genome size is expressed as 1C, or the amount of DNA in a gamete. As such, genome size estimation is a useful first step in a complete genome sequencing project. The number of sequencing reads required to produce a given level of coverage depends directly upon the 1C amount of DNA per cell, while an even more critical need is an accurate 1C genome size estimate to compare against the final assembly. As new technologies provide ever longer reads, the proportion of the genome assembled will approach the estimated genome size. However, large portions of the genome contain repeat elements, and these can vary in length. That variation means it will always be necessary to compare the assembled genome with the estimated size. While an accurate genome size estimate gauges the proportion of the genome assembled, an assembly that is greater than the estimate is a strong indication that structural variations, such as heterozygous inversions, have been assembled in

Susan J. Brown and Michael E. Pfrender (eds.), *Insect Genomics: Methods and Protocols*, Methods in Molecular Biology, vol. 1858, https://doi.org/10.1007/978-1-4939-8775-7_2,

tandem, rather than as alternative polymorphic sequences, and the tandem duplications have inflated the genome assembly.

Precision (estimates with small standard errors) is a necessity for intraspecific comparisons. Precise estimates of genome size by flow cytometry can help elucidate genome structure. Polyploidy, aneuploidy, and structural variants are revealed in comparisons within and between samples [1]. Cryptic species are often easily recognized by differences in DNA content [2]. Specialized DNA replication events, such as endopolyploidy [3, 4], chromatin diminution, or underreplication [5] can be discovered by comparing different tissues. Chromatin structure (euchromatin vs. heterochromatin) may be quantified by appropriate estimates of insect genome size [5, 6]. Differences between the genome size of males and females of a species provide feedback on the sex determination systems [7] and the role of endopolyploidy in sexual dimorphism [8]. The relative age of a "Y" (or w) sex chromosome is reflected by the extent of degeneration of sequences over time [9]. That degeneration is reflected by the difference in size of the genome of the homogametic and heterogametic sexes. A neo-X/Y is suggested when the genome size of heterogametic sex exceeds that of the homogametic sex. In a haplo/diplo species, a 2:1 ratio of genome size is seen in male/female comparisons. The presence of both a haploid peak and a second diploid peak in adult hymenopteran males revealed that acquired diploidy occurs in many, but not all adult males [10].

Intraspecific genome size variation exists and again the precision of estimates is an important consideration. Different strains and different inbred lines can differ significantly in genome size [11], and these differences may influence the choice of material for genome studies. Genome size variation between strains may also reflect life history differences [12, 13] that have important implications for genome size evolution.

2 Materials

1. Galbraith's buffer (*see* **Note 1**): 45 mM $MgCl_2$, 30 mM sodium citrate, 20 mM MOPS, 0.10% v/v Triton X, pH 7.0. Place a stir bar in a 1 L beaker on a stir plate. Add 450 mL distilled H_2O, and while stirring slowly, add 4.4 g sodium citrate ($Na_2C_6H_5O_7{\cdot}2H_2O$), 2.1 g MOPS (3-[*N*-morpholino]-propane sulfonic acid), 2.13 g $MgCl_2{\cdot}6H_2O$, 0.5 mL Triton X-100, and 50 μL 10 mg/mL RNAse A (*see* **Note 2**). Bring the total volume up to 500 mL with dH_2O. Slowly add HCl solution to lower the pH to 7.2 (*see* **Note 3**). Filter to sterilize and store in a sterilized container.
2. Propidium iodide (PI) (*see* **Note 4**) 1 mg/mL stock solution: Add 1 mg PI to 1 mL of Galbraith's buffer, or, if purchasing

100 mg quantities of PI make 100 mL and store in smaller aliquots at −80 °C.

3. 2 mL Dounce tissue grinder set with (required) type A (loose fitting) and (rarely required) B (tight fitting) pestles.
4. Pipet filter: Cut the ends off 2 1 mL pipet tips. Place 41 μm nylon mesh between pipet tips.

3 Methods

3.1 Tissue Preparation

1. Dissect live or fresh frozen tissue to be analyzed (*see* **Note 5**). Standards and unknowns should be prepared, stained, and measured together in the same tube (*see* **Note 6**).
2. Place tissue in 1 mL of ice-cold Galbraith's buffer in a 2 mL Dounce tissue grinder. Keep on crushed ice prior to grinding.
3. Grind the tissue with 15 gentle strokes using the A (loose) pestle (~3 strokes per 2 s) (*see* **Note 7**).
4. Filter through pipet tips containing nylon mesh into a microcentrifuge ("bullet") tube. Store on crushed ice.

3.2 Staining

1. Add Galbraith's buffer as needed to bring samples to 1 mL, add 25 μL PI stock and mix by inverting and gently shaking the tubes.
2. Store samples in the dark on ice or refrigerate for at least 20 min before running on the flow cytometer. Because the sample and standard may saturate at different rates, precision increases if all samples are run after the same period of staining; accuracy requires that saturation rates be compared by running the sample and standards over increasing stain time, working out the stain time when genome size reaches a plateau (Fig. 1). Some material may only plateau after samples stain for up to 24 h (*see* **Note 8**).

3.3 Running the Samples

1. Set cytometer to activate (trigger) on fluorescence, not scatter. Even with fluorescence activation, a tighter cleaner peak will be observed using a gate that excludes anything with high scatter (Fig. 2a–c). This gate will limit the output to fluorescence of clean intact nuclei. Because the goal is to determine the mean fluorescence from tight, symmetric G1 peaks from the sample and standard, a linear scale is essential (*see* **Note 9**). Log scales do not work for genome size estimation.
2. Set the flow rate to the slowest possible setting. This ensures that the nuclei are aligned in the center of the activating laser, minimizing the width of the fluorescence peaks. The slow rate will mean that fluorescence can be monitored over time. A second gate, based on red fluorescence over time, should be

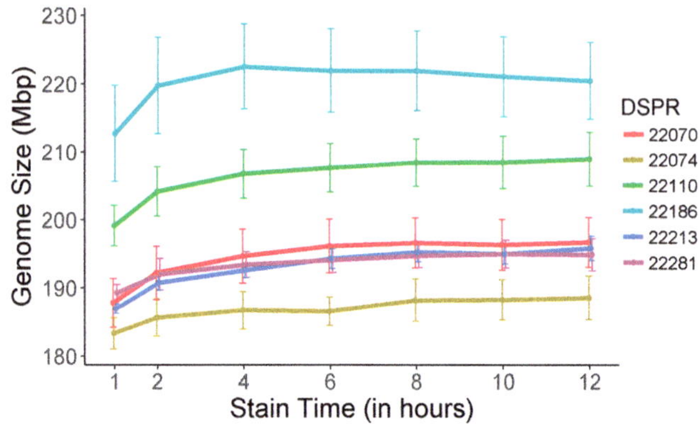

Fig. 1 Estimated genome size as a function of stain time. The estimated genome size will plateau at the most accurate value when the sample and standard reach the same levels of saturation. As shown here, optimal stain time can vary not only between species, but also among strains. Here, sample (DSPR strains of *Drosophila melanogaster*) and standard (*D. virilis* 1C = 328 Mbp) reach equal saturation levels after 4–8 h of stain time

used to select only the period when fluorescence output is constant (Fig. 2d). A tight, narrow peak is the best indication that the preparations are free of cytological tags from broken nuclei and other cell debris, and that the nuclei are aligned directly in the path of the exciting light. A tight peak is also critical to reduce bias due to overlap of peaks from the sample and standard (*see* **Note 9**).

3.4 Calculating Genome Size

Genome size determination is based on the relative position of the G1 (unreplicated) peaks of the sample and standard. It is critical that the ploidy of the first real (G1) fluorescent peak be correctly determined. To ensure this is so, the voltage to the red fluorescence photomultiplier tubes (PMT) must be adjusted with care. Genome size can vary between individuals by several orders of magnitude, and even within a single tissue preparation, highly polyploid nuclei may exist that produce a clean peak that is 2, 4, 8, or even 1200 times greater than that of the first G1 (diploid or haploid) nuclei (Fig. 3). In order to ensure the G1 peak is being estimated, the channel position of the red fluorescence should be adjusted upward until counts accumulate in the lowest channels, indicating unstained debris. Then the voltage should be decreased until the first clean tight peak is observed. This is the G1 peak that needs to be estimated. The red fluorescent peak position can then be adjusted for accurate estimation (channels 200–800 on many instruments). With a new sample, it is often a good idea to set up the flow cytometer with an easily prepared standard, such as *D. melanogaster*, and follow that with a sample (if sufficient material is available) of the unknown by itself. This allows you to determine

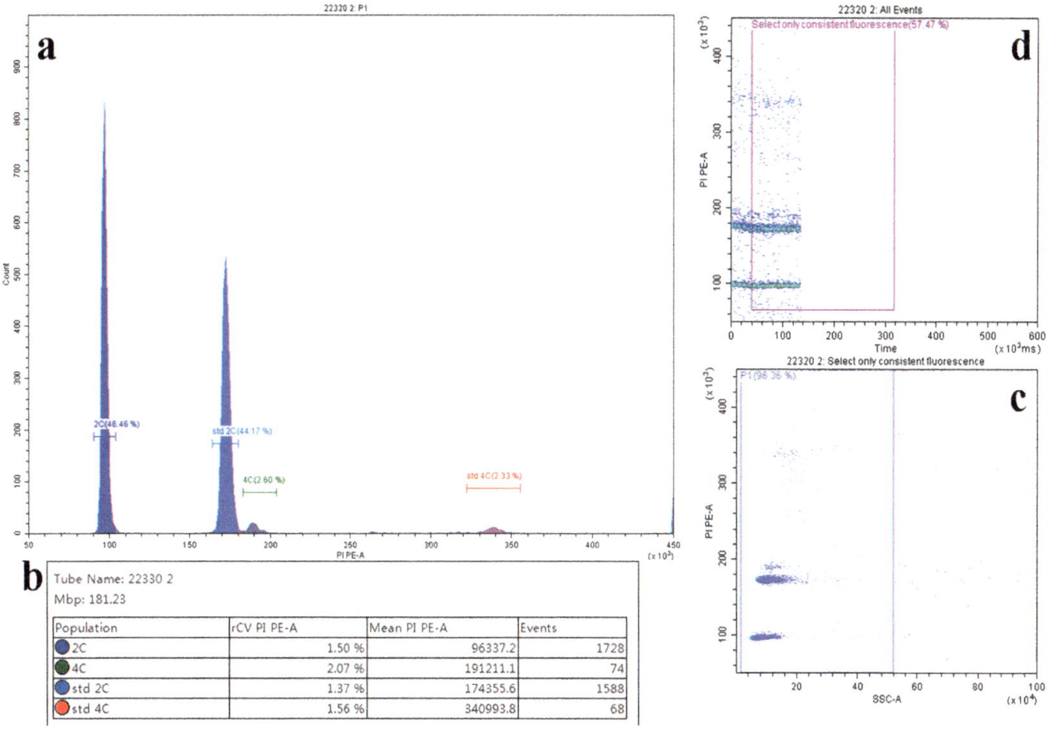

Population	rCV PI PE-A	Mean PI PE-A	Events
2C	1.50 %	96337.2	1728
4C	2.07 %	191211.1	74
std 2C	1.37 %	174355.6	1588
std 4C	1.56 %	340993.8	68

Fig. 2 Real-time flow cytometric histogram and cytograms. The G1 peaks of the sample and standard based on relative red propidium iodide fluorescence from PI-stained intact G1 nuclei is displayed as a histogram (**a**). The mean channel number of the G1 peaks, the number of nuclei scored under each peak, and the coefficient of variation for each peak is output (**b**) by internal flow cytometry software based on statistical gates set by the operator. A scatter gate (**c**) is set to exclude material with high scatter based on a cytogram showing red fluorescence versus side or forward scatter. A second time gate includes only unchanging fluorescence levels (**d**). To ensure that the histogram peaks are based on clean isolated nuclei free of cytological tags and free of fluorescence variation over time, the histogram is built from fluorescence output that meets criteria set by both scatter and time gates

the instrument settings for the sample, and allows you to choose the appropriate standard (discussed in the following paragraph). While the positions of separately run sample and standard peaks are important for selection of a standard and instrument settings, they should not be used for estimation of genome size.

3.5 Statistical Analysis

Methods have been devised to compensate for inhibitors and nonlinear responses. Among the most complete of these is described in [14]. These methods are most useful for plant material where multiple compounds, anthocyanin, caffeine, and tannins inhibit staining. Except for gall forming wasps that develop in the heavy tannins of oak galls, the methods are excessive for insect estimates. In general, the statistical gates built into the cytometer software are adequate to find the mean or peak position. Ideally the peaks are symmetric and the peak average channel number is the parameter of

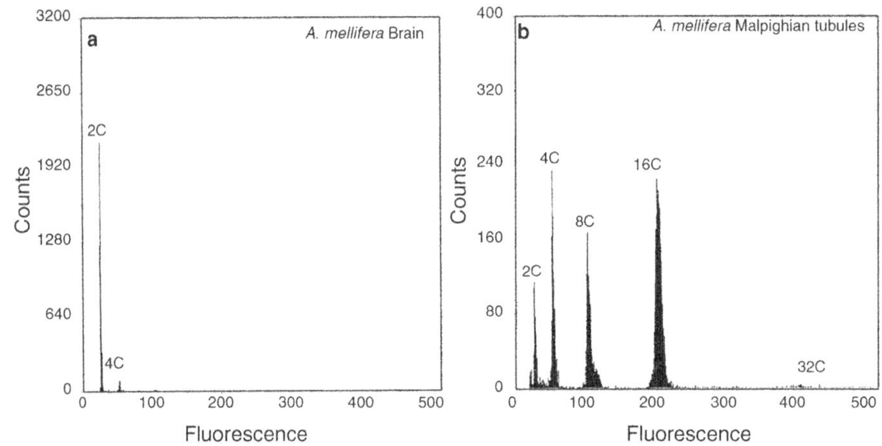

Fig. 3 Peaks produced by differing ploidy levels. Peaks produced by nuclei from G1, G2, and higher levels of endopolyploidy in the brain and Malpighian tubules of the worker honey bee, *Apis melifera*. Tissue suitable for genome size estimation will have a single strong symmetric G1 peak (**a**). Identification of the G1 peak may be problematic for tissues with nuclei at high ploidy levels (**b**)

choice to produce the lowest overall standard error about the mean. If the peak is skewed, the statistical gates can be set tighter around the peak, or the peak position itself may be used to provide a lower standard error. Second party software is available to fit a normal distribution about peaks. However best practice cytometry is preferred to methods that recover values from skewed or kurtotic peaks. Once the G1 peaks for the sample and standard have been identified and the mean or peak position scored, the amount of DNA in the unknown is a simple ratio of the average peak position of the unknown divided by the peak position of the standard times the 1C amount of DNA in the standard.

Care must be taken to interpret the genome size expressed as 1C. For the homogametic sex (X/X or Z/Z) the 1C value represents the amount of DNA in a gamete. For the heterogametic sex (X/O, X/Y, Z/O, Z/Z, Xi/Xj/Y, . . .) the 1C is the average of the two gametes produced. To find the difference in genome size associated with the sex chromosome, it is necessary to double the estimate then subtract. For example, to find the difference in size of the X and Y, where A represents the autosomal chromosome size common to both sexes, $X - Y = 2A + XX - 2A + XY$. Similarly, the size of the X can be found in an X/O system as $2A + XX - 2A + X/O = X$.

The CV (determined as the width at half the peak height or mean/variance) is useful to report the quality of the preparation and cytometry, but not useful for statistics. Calculation of average and standard errors using the PIVOT Table in MS Excel, or similar program is encouraged, as it is easy to recalculate estimates as new data are added, and easy to identify outlier values that inflate the

SE. Replication has the advantage that it allows reporting of the mean and the standard error of the mean (*see* **Note 10**). An average also solves another problem. A single genome estimate is a ratio of the peak position of the G1 nuclei from the sample and the standard. As such, it is a Cauchy distribution with no defined mean and variance. The central limit theorem ensures us that the average will be more normally distributed, thereby reducing problems with assumption underlying statistical tests.

The methods described here are useful for flow cytometric genome size estimates based on live or fresh frozen insects. Alternative methods that have been used to estimate genome size have been discussed elsewhere [15, 16]. The method described here was first described by Hare and Johnston 2011 [17]. The stain amount has been adjusted, additional information is provided for the different steps, and suggestions are added in an effort to provide greater accuracy as needed to determine the percent completion of assembled genomes.

4 Notes

1. Galbraith's buffer has the advantage that samples are stable over time, and cytometry permitting, allows technical replicates over time. Other buffers that have been recommended for problematic plant material [18] work well for insects and may produce excellent results with shorter stain times.
2. RNAse A can be purchased ready-to-use or as powder. If purchased as powder, it is important to boil the final stock solution 10 min to eliminate any DNase that might contaminate the product.
3. A pH slightly above 7.2 is ok, but not below 7.2. If the pH is too low, add NaOH to raise it to 7.2.
4. Propidium iodide is recommended as the stain, even though it stains both DNA and RNA and is actively pumped from living cells. Propidium iodide is recommended because it is stochastic (increases linearly with DNA amount), relatively insensitive to methylation and compaction of DNA, and produces an unbiased estimate, provided RNA has been removed by RNase. Other stains are available, such as DAPI and Geimsa that are specific for DNA and stain the nuclear DNA in whole cells. However, when used alone these stains produce biased estimates. The stains are sequence specific, which means estimates based on fluorescence reflect not only the actual genome size but also the AT/GC ratio. These stains also require shorter or longer wavelength excitation that is often not available on basic instruments. When instruments are available to score both PI and one or more of the sequence-specific stains, the

comparison of the estimates may be of interest, as an indication of the AT/GC ratio associated with specific repeats [19]. That information has been used to quantify intraspecific variation associated with variation in the repeats in sex chromosomes, telomere/centromere repeats, and highly repetitive sequences in supernumerary or "B" chromosomes.

5. The choice of tissue is often an easy one. The adult head of an adult insect usually has one strong peak of G1 nuclei (Fig. 3a). That peak is primarily produced by nuclei isolated from the soft, neural tissue. There are exceptions however, such as the highly polyploid tissues of the Malpighian tubules (Fig. 3b). Some biting insects, such as sandflies, have potent DNase in the salivary glands that will degrade the DNA in the co-prepared nuclei of the standard and the unknown. Careful dissection of neural tissue is required to create good peaks. Dung beetles may also have DNase that degrades DNA. Thorough washing and careful dissection of tissue without contact with the cuticle may help there. In a few insects, such as the cicada, stained nuclei isolated from the brain do not produce a tight peak. Fortunately, a single cicada eye is adequate to produce nuclei for a strong clean G1 peak. Two other examples where the head contained no suitable tissues for genome size estimates are pseudoscorpion (a chelicerate) and shrimp (a crustacean). The pseudoscorpion produces a clean G1 peak from the pedipalps and not from any other tissue. For shrimp, a small segment of a leg produces a strong G1 peak; other tissues fail to do so. Given that a peak of high quality and a suitable quantity of G1 nuclei cannot be produced with many tissues, it is always wise to try several tissues and identify not only ones that produce a strong tight peak, but also ones that produce a single strong peak from G1 nuclei.

6. Selection of the species to be used as the standard is very critical for co-preparation with the unknown. The standard should have a well-established genome size and produce one strong peak, with little background and few secondary peaks that may overlap the peak from the unknown. Ideally, the standard should fall between the 2C and 4C peaks of the unknown. More often, one must settle for a standard that is between ½X and 2X the size of the unknown. A standard chosen for its phylogenetic position relative to the unknown is likely to lead to bias, unless the genome size of that standard has been determined by a large number of replicates against another well-established standard from a model organism. The estimated genome size of the standard is one of the largest sources of bias in genome size estimates. Model insect species are recommended for standards for most comparisons [20]. The male of the seed beetle, *Callosobruchus maculatus* 1.23 pg.

(1C = 1205 Mbp), available from Carolina Biological Supply is an excellent alternative to chicken red blood cells (CRBCs) that are commonly used for estimates of genome size in that range. CRBCs will not saturate at the same rate as an insect cell, so a saturation curve must be run with the co-prepared sample and CRBC. Care must be taken as well to find the correct genome size of the CRBCs that are used. The literature includes values (2C) of 2.54 pg. (1C = 1240 Mbp), 2.78 pg. (1C =1360 pg.), and 2.33 pg. (1C = 1140 Mbp) for CRBCs. If the CRBC's are from a hen, these same values can be 0.04 pg. (1C = 20 Mbp) less. Bennett et al. found that 2.33 pg. [21], as originally estimated by Galbraith for nuclei extracted from a white leghorn rooster [22], provides values that agree well with the estimated values for model insect species.

7. Tissue grinding is an art, but one that is easily mastered. The goal of grinding is to release clean nuclei in numbers needed, without excessive debris. Grinding too fast or using too many strokes produces background from DNA released from and caught within broken nuclei. Too few strokes or grindingly too slowly releases few nuclei with subsequent long run times to accumulate 1000+ nuclei under Gl peaks. While fifteen strokes at a rate of three strokes every 2 s is generally a good starting point, soft bodied insects such as aphids may need fewer strokes, while hard bodied ones, including many of the beetles, may require slower grinding speeds.
8. The recommendation of 20 min of stain time is a minimum and assumes that the rate of stain uptake is the same for the sample and standard, which is not always true. In mosquitos with relatively large genomes, such as *Aedes*, the chromatin saturates slowly. Estimates taken after 20 min, 1 h, and 4 h will show the genome of the mosquito increasing by 10% or more. To ensure saturation of stain in the sample and standard, it is best to score the co-preparation after they have stained for different periods of time. The estimated genome size will eventually stabilize; it may be necessary to stain overnight to ensure equal stain saturation in the sample and standard.
9. The most challenging step for an experienced flow cytometer technician is the unavoidable variation in sample preparation for genome size estimates. Genome size estimation preparations are relatively crude and even the most careful grinding and filtering will result in broken nuclei, nuclei with adhering bits of DNA from broken nuclei and debris from the cell. All of these will scatter light and the very low scatter of a cleanly isolated nucleus is the best selection criteria for use in genome size estimates. The cytometer itself is a very important issue in genome size estimates. The cytometer should have a single excitation source in blue (488 nm) or green (514–533 nm)

and two parameter output, for red fluorescence and forward or side scatter. Most importantly, the cytometer must be capable of producing a peak from a stained nuclear preparation from a *Drosophila melanogaster* yellow body, white eye strain (y/w) that is symmetric with a coefficient of variation (CV) of 2.0 or less. Overlap of sample and standard peaks, which is inevitable with broad peaks, will bias the estimate. The red fluorescent output from the cytometer must also be linear. All manufacturers claim their cytometers are linear, yet nearly all have an offset from zero in order to produce a symmetric peak for unstained nuclei. The best way to demonstrate linearity is to run sufficient numbers of nuclei to create reliable 2C and 4C peaks. If the 4C is not twice the 2C, linearity is a problem. Running samples and standards between channel 200 and 800 will reduce, but not entirely eliminate bias due to nonlinearity. Running samples and standards that are close to one another, but not overlapping, also helps. Clean tight peaks are even more critical in tests for small intraspecific differences. The critical test for real differences is to co-prepare samples that were scored as different when prepared individually. If the differences are artifacts of preparation [23], the co-preparation will produce a single peak with a CV that is almost as low as that of either sample alone (Fig. 2b). In contrast, real genome size differences, whether between sexes or between individuals of the same sex reared in different locations and different environmental conditions, will be revealed by the presence of two side-by-side florescence peaks.

10. The offset built into the cytometer is not always the same each day and that can produce a significant day effect. The only easy solution is to run replicates of an unknown on different days. Lacking that, it is possible to produce an estimate that is very precise, but not accurate. Two types of replicates are generally possible and always desirable with cytometry. The sample may be split into two or three aliquots and each run as a technical replicate. These are most useful to determine the optimal stain time, but will also provide feedback on machine (run-time) error. Different preparations from different tissues will provide feedback on errors due to preparation and tissue type and will help determine the optimal tissue for use in in biological replicates. The more important replication is biological. Whenever possible, co-preparations of five samples plus the standard should be prepared and run on different days. Biological replication has the advantage that it allows reporting of the average genome size and the standard error.

References

1. Jacobson AL, Johnston JS, Rotenberg D, Whitfield AE, Booth W, Vargo EL, Kennedy GG (2013) Genome size and ploidy of thysanoptera. Insect Mol Biol 22(1):12–17. https://doi.org/10.1111/j.1365-2583.2012.01165.x
2. Lower SS, Spencer Johnston J, Stanger-Hall K, Hjelmen CE, Hanrahan SJ, Korunes K, Hall D (2017) Genome size in North American fireflies: substantial variation likely driven by neutral processes. Genome Biol Evol 9 (6):1499–1512. https://doi.org/10.1093/gbe/evx097
3. Rangel J, Strauss K, Seedorf K, Hjelmen C, Johnston J (2015) Endopolyploidy changes with age-related polyethism in the honey bee, *Apis mellifera*. PLoS One 10(4):e0122208. https://doi.org/10.1371/journal.pone.0122208
4. Scholes DR, Suarez AV, Paige KN (2013) Can endopolyploidy explain body size variation within and between castes in ants? Ecol Evol 3 (7):2128–2137
5. Johnston JS, Schoener M, McMahon DP (2013) DNA underreplication in the majority of nuclei in the *Drosophila melanogaster* thorax: evidence from Suur and flow cytometry. J Mol Biol Res 3(1):47
6. Bosco G, Campbell P, Leiva-Neto JT, Markow TA (2007) Analysis of *Drosophila* species genome size and satellite DNA content reveals significant differences among strains as well as between species. Genetics 177(3):1277–1290
7. Barcenas N, Thompson N, Gomez-Tovar V, Morales-Ramos J, Johnston J (2008) Sex determination and genome size in Catolaccus grandis (Burks, 1954) (Hymenoptera: Pteromalidae). J Hymenopt Res 17:201–209
8. Johnston J, Ross L, Beani L, Hughes D, Kathirithamby J (2004) Tiny genomes and endoreduplication in Strepsiptera. Insect Mol Biol 13(6):581–585
9. Bachtrog D (2013) Y-chromosome evolution: emerging insights into processes of Y-chromosome degeneration. Nat Rev Genet 14(2):113–124
10. Aron S, de Menten L, Van Bockstaele DR, Blank SM, Roisin Y (2005) When hymenopteran males reinvented diploidy. Curr Biol 15 (9):824–827
11. Huang W, Massouras A, Inoue Y, Peiffer J, Ramia M, Tarone AM, Turlapati L, Zichner T, Zhu D, Lyman RF, Magwire MM, Blankenburg K, Carbone MA, Chang K, Ellis LL, Fernandez S, Han Y, Highnam G, Hjelmen CE, Jack JR, Javaid M, Jayaseelan J, Kalra D, Lee S, Lewis L, Munidasa M, Ongeri F, Patel S, Perales L, Perez A, Pu L, Rollmann SM, Ruth R, Saada N, Warner C, Williams A, Wu YQ, Yamamoto A, Zhang Y, Zhu Y, Anholt RRH, Korbel JO, Mittelman D, Muzny DM, Gibbs RA, Barbadilla A, Johnston JS, Stone EA, Richards S, Deplancke B, Mackay TFC (2014) Natural variation in genome architecture among 205 *Drosophila melanogaster* genetic reference panel lines. Genome Res 24 (7):1193–1208. https://doi.org/10.1101/gr.171546.113
12. Arnqvist G, Sayadi A, Immonen E, Hotzy C, Rankin D, Tuda M, Hjelmen CE, Johnston JS (2015) Genome size correlates with reproductive fitness in seed beetles. Proc R Soc B 282 (1815):20151421
13. Ellis LL, Huang W, Quinn AM, Ahuja A, Alfrejd B, Gomez FE, Hjelmen CE, Moore KL, Mackay TFC, Johnston JS, Tarone AM (2014) Intrapopulation genome size in *D. melanogaster* reflects life history variation and plasticity. PLoS Genet 10(7):e1004522. https://doi.org/10.1371/journal.pgen.1004522
14. Barre P, Noirot M, Louarn J, Duperray C, Hamon S (1996) Reliable flow cytometric estimation of nuclear DNA content in coffee trees. Cytometry 24(1):32–38
15. Desalle R, Gregory TR, Johnston JS (2005) Preparation of samples for comparative studies of arthropod chromosomes: visualization, in situ hybridization, and genome size estimation. Methods Enzymol 395:460–488
16. Gregory TR, Nathwani P, Bonnett TR, Huber DP (2013) Sizing up arthropod genomes: an evaluation of the impact of environmental variation on genome size estimates by flow cytometry and the use of qPCR as a method of estimation. Genome 56(9):505–510
17. Hare EE, Johnston JS (2011) Genome size determination using flow cytometry of propidium iodide-stained nuclei. In: Orgogozo V, Rockman MV (eds) Molecular Methods for Evolutionary Genetics, vol 772. Humana Press, New York, NY, pp 3–12. https://doi.org/10.1007/978-1-61779-228-1_1
18. Loureiro J, Rodriguez E, Doležel J, Santos C (2006) Flow cytometric and microscopic analysis of the effect of tannic acid on plant nuclei and estimation of DNA content. Ann Bot 98 (3):515–527
19. Šmarda P, Bureš P, Šmerda J, Horová L (2012) Measurements of genomic GC content in plant

genomes with flow cytometry: a test for reliability. New Phytol 193(2):513–521

20. Hanrahan SJ, Johnston JS (2011) New genome size estimates of 134 species of arthropods. Chromosom Res 19:809–823. https://doi.org/10.1007/s10577-011-9231-6
21. Bennett MD, Leitch IJ, Price HJ, Johnston JS (2003) Comparisons with Caenorhabditis (~100 Mb) and Drosophila (~175 Mb) using flow cytometry show genome size in Arabidopsis to be ~157 Mb and thus ~25% larger than the Arabidopsis genome initiative estimate of ~125 Mb. Ann Bot 91(5):547–557
22. Galbraith DW, Harkins KR, Maddox JM, Ayres NM, Sharma DP, Firoozabady E (1983) Rapid flow cytometric analysis of the cell cycle in intact plant tissues. Science 220(4601):1049–1051
23. Bennett MD, Price HJ, Johnston JS (2007) Anthocyanin inhibits propidium iodide DNA fluorescence in Euphorbia pulcherrima: implications for genome size variation and flow cytometry. Ann Bot 101(6):777–790

Chapter 3

Isolation of High Molecular Weight DNA from Insects

Susan J. Brown and Michelle Coleman

Abstract

Isolation of high molecular weight (HMW) DNA from any organism can be daunting and populations of extremely small insects with hard exoskeletons are no exception. It can be challenging to isolate DNA that is clean enough to enzymatically label, digest or otherwise manipulate without sacrificing length. The method described here was originally developed to isolate HMW DNA for BAC library construction. More recently, it has proven reliable for long range optical mapping as well.

Key word High molecular weight DNA, Optical mapping, Long range restriction mapping

1 Introduction

The quality of a genome assembly depends, to a large extent, on depth of coverage and length of sequencing reads. Whole genome shotgun sequencing depends on the ability to generate long contiguous stretches of DNA (contigs) from relatively short sequencing reads. Repetitive DNA, which can comprise a large part of an insect genome, impedes genome assembly, resulting in higher fragmentation and lower scaffold N50. While usually serviceable for gene finding, such assemblies provide little insight into genome organization and may even misinform gene structure. Sequencing and mapping technologies that are more likely to span a repeat and provide a glimpse into unique flanking DNA sequence require longer input DNA molecules [1].

Most next generation sequencing technologies require starting material around 50 kb to construct short insert libraries for contig construction and long insert libraries for scaffolding. High molecular weight (HMW) DNA between 50 and 250 kb is needed for newer methods that produce contigs based on long reads. Long range sequencing methods and optical mapping require even longer starting material to generate libraries of 100–350 kb DNA, similar to traditional BAC libraries [2]. Isolation kits are available to recover DNA of various lengths, depending on specific needs/

Susan J. Brown and Michael E. Pfrender (eds.), *Insect Genomics: Methods and Protocols*, Methods in Molecular Biology, vol. 1858, https://doi.org/10.1007/978-1-4939-8775-7_3, © Springer Science+Business Media, LLC, part of Springer Nature 2019

approaches. While these provide clean DNA of excellent quality and length for short-to-midrange insert reads, and they may recover a little DNA up to 1 Mb, the amount of very HMW DNA >250 kb is usually insufficient for long range sequencing or optical mapping analysis. Here we describe a method to isolate very HMW DNA from insects that is based on our experience with *Tribolium castaneum* [3].

2 Materials

2.1 Nuclear Isolation

Liquid nitrogen.

Mortar and pestle (stored in −80 °C).

100 μm cell strainer.

40 μm cell strainer.

Glass centrifuge tubes.

Small funnel to fit glass tubes.

Wide bore 1000 μL tips.

Metal spoon type spatula.

Ice bucket.

Cryo-gloves.

Lowmelt agarose: 500 μL aliquots stored at 4 °C (*see* **Note 1**).

85 μL Plug Molds (*see* **Note 2**).

Aluminum block.

Refrigerated centrifuge (4 °C) with swinging bucket rotor.

Water or bead baths set to 43 °C and 70 °C for agarose and nuclei suspension warming.

NIB: Nuclear Isolation Buffer (store at 4 °C for up to a month) 10 mM Tris pH 9.4, 60 mM NaCl, 10 mM EDTA, 0.15 mM Spermidine trihydrochloride, 0.15 mM Spermine tetrahydrochloride, 0.5% Triton X-100, Beta-mercaptoethanol (BME). To make 200 mLs, mix 2 mL of 1 M Tris pH 9.4, 2.4 mL of 5 M NaCl, 4.0 mL of 0.5 M EDTA 8.0, 0.3 mL 0.1 M Spermidine (stored at −20 °C), 0.3 mL 0.1 M Spermine (stored at −20 °C), 1 mL Triton X 100 with 190 mL of ultrapure water.

2.2 Nuclear Lysis

50 mL conical tubes.

Green screened cap for 50 mL conical tubes.

Proteinase K (*see* **Note 3**).

Lysis Buffer (1% SDS, 100 mM Tris–HCL pH 8.0, 50 mM EDTA pH 8.0, and 100 mM NaCl).

2.3 DNA Isolation

RNAse A (*see* **Note 3**).

High EDTA buffer pH 8.0. To make 1 L (10 mM Tris–HCL pH 8.0, 50 mM EDTA pH 8.0), mix 10 mL of 1 M Tris–HCL pH 8.0, 100 mL of 0.5 M EDTA pH 8.0, to 890 mL of water. Store at room temperature.

3 Methods

Work efficiently through each step and keep sample on ice at all times. To recover HMW DNA, gentle handling, kept to a minimum, is key to integrity of DNA. Use wide bore tips and pipette very slowly, do not vortex, shake, or drop tubes.

3.1 Nuclei Isolation

1. Melt a 500 μL aliquot of 2% LMA at 70 °C for 10 min. Equilibrate melted agarose at 43 °C for at least 30 min to prior to embedding.
2. Immediately prior to grinding, remove small mortar and pestle from −80 °C freezer. Fill mortar (containing pestle) with liquid nitrogen and wait for bubbling to stop. Refill as needed. Wear cryo-gloves to project hands. Add 0.5 g of sample (*see* **Note 4**) to mortar half full of liquid nitrogen and grind until a fine powder is produced (~30 s to a minute). Make sure sample is always submerged in the liquid nitrogen (*see* **Note 5**).
3. After liquid nitrogen has sublimated, using round spatula scraper, collect powder near spout of mortar. Carefully scrape powder into glass tube (on ice) containing 10 mL of chilled NIB (*see* **Note 6**). Mix powder into NIB with round end spatula. Wet powder completely.
4. Immediately pour resuspended powder through 100 μm cell strainer (yellow) into glass centrifuge tube (on ice) (*see* **Note 7**).
5. Spin at 100 × *g* at 4 °C for 15 s (counted after rotor is up to speed) to pellet cellular debris (pellet will be beige). Carefully pour supernatant (containing nuclei) into a clean cold glass centrifuge tube. Spin at 1800 × *g*, 4 °C for 3 min to pellet nuclei (*see* **Note 8**).
6. Tip tube to pour off supernatant, being careful to **retain pellet** (it will be loose). While holding tube upside down, touch lip of tube to a clean paper towel to remove last of supernatant.
7. Cover pellet with 1 mL ice-cold NIB. Gently resuspend pellet with 1 mL wide bore tip. Add NIB bringing volume to 20 mL.
8. Spin at 1800 × *g*, 4 °C for 3 min to pellet nuclei. Pellet should be **whitish**, not beige or yellow. When resuspended, solution should be translucently whitish. (*see* **Note 9**). Repeat **steps 6–8** if needed.

9. After final wash, cover pellet with cell suspension buffer up to 500 μL (*see* **Note 10**). Gently resuspend nuclei with wide bore tip and move resuspended nuclei to clean 1.5 mL tube.

3.2 Nuclei Embedding and Nuclei Lysis

1. Move plug mold from 4 °C to ice bucket immediately before use.
2. Move 1.5 mL tube containing nuclei from ice to 43 °C. Equilibrate for 4 min.
3. Using a wide bore tip, add 2% LMA to cells to achieve a final concentration of 0.7% agarose per plug. With a wide bore tip, pipet ~5 times to mix agarose and nuclei mixture completely.
4. Begin filling plug molds (85 μL per plug) immediately (*see* **Note 11**). Mix the agarose/nuclei suspension three times with a wide bore tip between filling each plug mold (*see* **Note 12**).
5. Place plug mold in 4 °C cooler until LMA/nuclei plugs solidify, approximately 20 min.
6. Push plugs out of mold into a 50 mL conical tube containing 2.5 mL lysis buffer and 200 μL Puregene Proteinase K (*see* **Note 13**). After 2 h strain buffer through green screen cap and add a fresh 2.5 mL aliquot of Buffer and Proteinase K to plugs. Incubate tubes overnight at 50 °C (*see* **Note 14**).
7. In the morning, remove tubes from 50 °C and let equilibrate to room temp for 10 min. Add 50 μL RNAse A and incubate at 37 °C for 1 h.

3.3 DNA Purification in Agarose Plus

1. Wash plugs for 15 min in 10 mL high EDTA buffer, repeat twice. HMW DNA is most stable in plugs at this stage. They can be shipped or stored for prolonged periods at 4 °C in high EDTA buffer (*see* **Note 15**).

4 Notes

1. High quality LMA as a 2% solution works best. Make several ml in ultrapure water and store in 500 μL aliquots at 4 °C.
2. Plug molds are available in many sizes, a volume of 85 μL provides best dimensions for reagent and enzyme diffusion.
3. High quality Proteinase K and RNAse A work best.
4. For insects, the least complex tissue is the best. Late stage embryos contain hundreds-thousands of nuclei, but have not completed organogenesis. Newly hatched larvae or pupa may also work. Adults and late larva are usually too complex. If possible, remove any food substrates from sample prior to grinding. The quality of the starting material is essential to

success. Live samples work best, flash freezing (−80 °C) of healthy insects can be tried.

5. Alternatively, you can use a mortar that has two pieces, a lower half (cooler) that holds liquid nitrogen and an upper half (metal bowl) that is removable that you also add liquid nitrogen to and do the grinding in to keep sample ultra-cold.
6. Before starting extraction, remove 1 mL of NIB (without BME) and store on ice. You will use up to 500 μL of this to resuspend final pellet of nuclei for embedding. Add 200 μL of fresh BME to remaining NIB (199 mL) prior to use. Once BME is added, the NIB solution must be used that day and cannot be stored! Only fresh BME functions as a potent reductant.
7. You may have to put a funnel in the tube and seat the cell strainer in the funnel if the cell strainer doesn't fit into the glass tube.
8. 1800 × *g* works well for a 200 Mb genome. If nuclei do not pellet or pellet is small, higher speed may be required to pellet nuclei containing smaller genomes.
9. Multiple washes (up to three times) may be needed to produce a clean white pellet. Be mindful, the pellet is loose, and nuclei will be lost in each wash.
10. We have found that pellet sizes vary. If the pellet is small, resuspend it in less than 500 μL NIB (no BME). Alternatively, if the pellet is really large, you may want to make a dilution series and make pellets of all dilutions into separate plugs. Please remember cell suspension buffer is the NIB aliquot without BME added.
11. To super chill plug mold and encourage solidification of agarose before nuclei have a chance to settle to bottom of plug, an aluminum block, pre-chilled at 4 °C, can be placed in the ice bucket and the plug mold placed on top of the block.
12. Work quickly but gently so that agarose does not harden prematurely and nuclei are not damaged pre-embedding.
13. Place 5 plugs per 50 mL tube max for maximum equilibration of reagents.
14. Intermittent gentle mixing during overnight incubation optimizes equilibration of reagents.
15. HMW DNA can also be manipulated with restriction enzymes directly in the agarose plugs. Alternatively, the agarose plugs may be enzymatically degraded and the HMW DNA recovered by membrane dialysis [4].

References

1. Zhang M, Zhang Y, Scheuring C, Wu C, Dong J, Zhang H (2012) Preparation of megabase-sized DNA from a variety of organisms using the nuclei method for advanced genomics research. Nat Protoc 7(3):467–478
2. Peterson DG, Tomkins JP, Frisch DA, Wing R, Paterson AH (2000) Construction of plant bacterial artificial chromosome (BAC) libraries: an illustrated guide. J Agric Genomics 5:1–100
3. Brown SJ, Henry JK, Black WC, Denell RE (1990) Molecular genetic manipulation of the red flour beetle: genome organization and cloning of a ribosomal protein gene. Insect Biochem 20(2):185–193
4. Marusyk R, Sergeant A (1980) A simple method for dialysis of small-volume samples. Anal Biochem 105:403–404

Chapter 4

Long Range Sequencing and Validation of Insect Genome Assemblies

Surya Saha

Abstract

Advances in long read and long range sequencing technologies have enabled chromosome length resolution for de novo genome assemblies even in the absence of complementary resources such as physical maps. Herein, I introduce a few methods for quality control and discuss potential pitfalls when assembling insect genomes with long reads.

Key words Assembly, Scaffolding, Next-generation sequencing, *De bruijn* graph, PacBio, Nanopore

1 Introduction

Numerous arthropod genome projects have originated out of the i5k [1] project that were started with second generation sequencing technology such as Roche 454 and Solexa. These were primarily based on short read platforms, so the quality of these assemblies varies from good to relatively poor draft genomes. High quality gene annotation requires assemblies to have a N50 of 250–350Kb or greater [2]. Given the diversity among insect species and the lack of reference genomes, phylogenetically close species with a high quality genome assembly may not be available to guide assembly and annotation for a new project. The prevalence of complex gene families like olfactory receptors and P450 genes, DNA polymorphisms and transposable elements in arthropod genomes [2–4] further obfuscates resolution of the assembly using only short reads.

The advent of third generation long read sequencing methods offers an effective option for generating high quality arthropod reference assemblies. The improved genome assembly of *Aedes aegypti*, the vector of Zika virus, exemplifies improvements that can be achieved when novel methods and technologies are applied. The original *Aedes aegypti* genome was assembled from 8X Sanger sequencing into 4756 scaffolds at an N50 of 1.5 Mb

Susan J. Brown and Michael E. Pfrender (eds.), *Insect Genomics: Methods and Protocols*, Methods in Molecular Biology, vol. 1858, https://doi.org/10.1007/978-1-4939-8775-7_4,

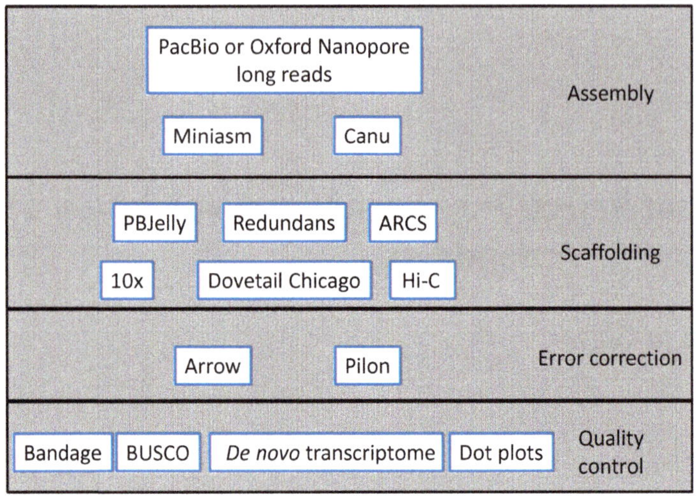

Fig. 1 Genome assembly and validation workflow

[5]. Dudchenko et al. [6] generated 40X Hi-C coverage that enabled them to order, orient, and merge scaffolds into three chromosomes. They also identified and corrected 1422 misassemblies in the previous assembly. Applying the Hi-C method to improve the genome of *Culex quinquefasciatus*, the vector for West Nile virus, they were able to generate a high quality assembly with three chromosome-length scaffolds.

The reader is referred to the chapter on genome sequencing and assembly strategies for an overview of the recent developments in sequencing technologies and types of data sets required for a high quality genome assembly. The primary goal of this chapter is to provide a workflow to generate a high quality genome assembly using long range sequencing technologies and evaluate it for common assembly errors using various metrics. An overview of the steps is presented in Fig. 1, and Table 1 contains information about the websites for all the tools utilized in the pipeline.

2 Materials

1. Fastq files from a sequencing facility (*see* **Note 1**).
2. A Unix or Linux computer to run the workflow commands (*see* **Note 2**).
3. Software (*see* Table 1).

Table 1
List of tools utilized in the workflow presented in this chapter

Software	Website	References
Canu assembler	http://canu.readthedocs.io/en/latest/index.html	[7]
Miniasm assembler	https://github.com/lh3/miniasm	[8]
PBJelly scaffolding tool	https://sourceforge.net/projects/pb-jelly/	[9]
ARCS scaffolding tool	https://github.com/bcgsc/arcs	[10]
Redundans reduction and scaffolding tool	https://github.com/Gabaldonlab/redundans	[11]
Arrow error correction	https://github.com/PacificBiosciences/GenomicConsensus	
PILON error correction	https://github.com/broadinstitute/pilon/wiki	[12]
BUSCO single copy gene markers	http://busco.ezlab.org/	[13]
Bandage graph assembly viewer	https://rrwick.github.io/Bandage/	[14]
Gepard dotter	http://cube.univie.ac.at/gepard	[15]
MUMmer aligner and plotter	http://mummer.sourceforge.net/	[16]

3 Methods

3.1 Assembly

1. Miniasm assembly
The Miniasm assembler [8] does not correct reads before assembly unlike most other long read assemblers. It has been designed to rapidly assemble reads from PacBio and Oxford Nanopore data to produce uncorrected contigs. It takes as input an all-to-all alignment from Minimap [17] and identifies all raw read overlaps. The mappings are then trimmed and modeled into an assembly graph before unitigs are determined from paths through the graph. We use Miniasm to rapidly generate candidate assemblies from uncorrected raw reads as well as corrected reads.
(a) Install git using the appropriate package manager, i.e., for Debian.

```
aptitude install git
```

(b) Clone the repository from github and build the executables.

```
git clone https://github.com/lh3/minimap &&
  (cd minimap && make)
```

git clone https://github.com/lh3/miniasm && (cd miniasm && make)

(c) Run the overlap step on a multi-core computer with 16 cores.

minimap/minimap -Sw5 -L100 -m0 –t64 reads.fq reads.fq | gzip -1 > reads.paf.gz

(d) Layout step.

miniasm/miniasm -f reads.fq reads.paf.gz > reads.gfa

(e) Create the fasta file of contigs from the .gfa file.
zcat reads.paf.gz | awk '$1=="S" {print ">"$2"\n"$3}' > reads.paf.gz.fa

2. Canu assembly
The Canu assembler [7] is an adaptation of the Celera assembler [18] that assembles long reads with high error rates. Canu starts by self-correcting the raw reads using a MinHash Alignment Process (MHAP) [19] to compare raw reads and compute overlaps. Unitigs are constructed from a sparse read overlap graph with a version of the best overlap graph algorithm [20] in which the local and global overlap error rates are estimated from the data itself. Use Canu to error correct, trim, and assemble PacBio data to create contiguous candidate genome assemblies (*see* **Note 3**).

(a) Clone the repository from github and build the executables.

git clone https://github.com/marbl/canu.git

cd canu/src

make -j 64

(b) Self-correct the raw long Pacbio reads using the shorter Pacbio reads. The command presumes that the genome size of the insect is 450 Mb.

canu/Linux-amd64/bin/canu -correct –p canu -d reads.fq-canu genomeSize=450m -useGrid=0 -maxThreads=60 -maxMemory=500 -pacbio-raw reads.fq

(c) Trim the corrected reads based on the overlap with other corrected reads so that erroneous regions of reads are excised.

canu/Linux-amd64/bin/canu -trim –p canu -d reads.fq-canu genomeSize=450m -useGrid=0 -maxThreads=60 -maxMemory=500 -pacbio-corrected reads.fq-canu /canu.correctedReads.fasta

(d) Finally assemble the contigs from the corrected and trimmed Pacbio reads.

```
canu/Linux-amd64/bin/canu -assemble -p canu -d
    reads.fq-canu-assembly genomeSize=450m
    -useGrid=0 -maxThreads=60 -maxMemory=500
    -pacbio-corrected reads.fq-canu /canu.trimmed-
    Reads.fastq
```

3. PBJelly scaffolding

One of the first applications of PacBio sequencing was to fill gaps and scaffold draft genome assemblies created with short reads from Illumina and 454 pyrosequencing platforms. Low coverage PacBio reads are mapped iteratively to a draft genome assembly using BLASR [21] by PBJelly [9]. This is repeated at varying overlap thresholds in order to identify contigs that are linked by multiple PacBio reads. The ends of the contigs are extended into the gap in cases where the PacBio read does not span the entire gap region. This can result in the closure of the gap in later iterations if enough support is found to join the extended contigs. PBJelly is used to scaffold the contigs from the Canu and Miniasm assemblies to connect unitigs and improve the contiguity of the assembly. The PBJelly output should be carefully examined to ensure that only valid connections are created between unitigs.

(a) Install PBJelly by following instructions at https://sourceforge.net/p/pb-jelly/wiki/Home/?#058c

(b) Modify the paths to dependencies, source the environment file and add the paths to input files in Protocol.xml. Initialize the PBJelly pipeline by tagging sequence names, finding gaps, and indexing the assembly.

```
emacs setup.sh
source setup.sh
Jelly.py setup Protocol.xml
```

(c) Use blasr to map the Pacbio sequences to the assembly.

```
Jelly.py mapping Protocol.xml
```

(d) Find support for the gaps in the assembly.

```
Jelly.py support Protocol.xml
```

(e) Identify Pacbio reads that extend into gaps or bridge gaps is possible.

```
Jelly.py extraction Protocol.xml
```

(f) Assemble the Pacbio reads in the gaps.

```
Jelly.py assembly Protocol.xml
```

(g) Finally create the output files for the scaffolded assembly. Jelly.py output Protocol.xml

4. Redundans duplication removal and scaffolding
   Highly heterozygous insect genome assemblies may contain a large number of duplicates even after careful genome assembly. Redundans [11] is a pipeline to remove duplicated contigs from an assembly, followed by iterative gap closing and scaffolding using paired-end, mate-pair or long reads.
   (a) Install Redundans from https://github.com/Gabaldonlab/redundans
   (b) Run the reduction step followed by scaffolding with Illumina paired-end and Pacbio long read data.
   redundans.py -v -i *.fq.gz -l pacbio.fq.gz -f assembly.fa -o test –t 32
5. Dovetail and 10X assembly correction and scaffolding
   Chicago libraries [22] utilize chromatin crosslinks to associate sequences originating from the same large DNA fragment. Linked reads generated by 10X from the same DNA molecule are tagged by a unique bar code [23]. An assembly that is created with input DNA from multiple individuals can introduce additional heterozygosity in the assembly. Chicago and 10X DNA sequences, ideally from a single individual (*see* **Note 1**), can be used to correct misassemblies by breaking chimeric contigs and also resolving collapsed repeats and fragmented polymorphic regions. 10X libraries can be used to join contigs into larger scaffolds with tools such as ARCS [10] while scaffolding with Chicago data can be performed with the HiRise assembler [22].

3.2 Error Correction Tools

Although the assembly process includes read correction during the assembly process, it is advisable to perform multiple rounds of error correction after the final assembly is created.

1. Arrow
   The Arrow algorithm from the GenomicConsensus package from PacBio (https://github.com/PacificBiosciences/GenomicConsensus) uses a hidden markov model based on the polymerase and chemistry used to generate the PacBio data for correction. PacBio reads are known to contain insertion and deletion errors [24] that can confound gene prediction besides impacting high resolution analysis like SNPs. Arrow also reports variants in the raw PacBio reads that can be helpful to identify alleles in the population if multiple individuals were used to generate the DNA sequence.
   (a) Install arrow using Bioconda (https://bioconda.github.io/) from https://github.com/PacificBiosciences/pbbioconda

(b) Install the GenomicConsensus package that contains Arrow.

```
conda install -c bioconda GenomicConsensus
```

(c) Create a file with list of all subreads.bam files.

```
find data/ -name "*subreads.bam" > subreads.bams.fofn
```

(d) Align subreads from each SMRT cell to the assembly and correct it using arrow iteratively.

```
while read LINE
do pbalign --verbose --nproc 32 $LINE assembly.fasta
    out_pacbio.bam
samtools faidx assembly.fasta
arrow --verbose -j 32 --annotateGFF --reportEffective-
    Coverage --referenceFilename assembly.fasta -o
    updated.fixes.gff -o updated.assembly.fasta -o
    updated.assembly.fastq out_pacbio.bam
mv updated.assembly.fasta assembly.fasta
done < subreads.bams.fofn
```

2. Pilon

Pilon [12] uses high quality Illumina sequences to polish PacBio and Oxford Nanopore long reads assemblies. Illumina reads are aligned to the genome assembly and supplied to Pilon as a BAM file [25]. Illumina reads from unpaired and paired-end libraries can be used as input. The correct base at a position is inferred by consensus and the reference allele is corrected if required. Pilon can also fix small insertion and deletion errors, fill gaps and perform local reassembly in regions where many errors are identified (*see* **Note 4**).

(a) Install pilon from https://github.com/broadinstitute/pilon/wiki and hisat2 from http://ccb.jhu.edu/software/hisat2/index.shtml

(b) Align Illumina reads to the assembly using hisat2

```
hisat2-2.0.5/hisat2 --time --threads 60 --mm --no-
    spliced-alignment -x assembly.fasta -U unpaired.fq
    2>ht2.sdtout | samtools view -@60 -hu -| samtools
    sort -@60 -O bam - > sorted.ht2.assembly.bam
```

(c) Run pilon to correct the assembly and create a detailed log of all the fixes

```
java -d64 -Xms512m -Xmx450g -jar pilon-1.22.jar --
    genome assembly.fasta --unpaired sorted.ht2.assem-
    bly.bam --threads 64 --changes --fix bases --diploid
    --output unpaired --outdir pilon-unpaired --debug
    3>&1 1>&2 2>&3 > unpaired-pilon.log
```

3.3 Quality Control

A genome assembly project typically involves an exploration of the parameter space to identify optimal parameters and combination of tools to generate the best assembly for a target genome, given the input data. Comparison of candidate genome assembles and developing meaningful metrics is an important step before an assembly can be carefully evaluated.

1. Bandage
   Bandage [14] is a visualization and analysis tool for exploring assembly graphs that are reported by assemblers such as Canu and Miniasm. It allows users to interact with an assembly and manually resolve contigs that contain errors. We use Bandage to compare the unitig graph to the contig graph when evaluating contigs with low support. Bandage includes an integrated Blast tool that is very useful to locate sequences and relevant nodes on the assembly graph.
2. Gepard and Mummer
   Pairwise sequence alignment of assembled contigs is very useful for comparing a scaffold to its constituent contigs. This is also effective for identifying structural variations among contigs from a single assembly or from multiple candidate assemblies. Dot plots aid in rapid visual analysis of high level structural comparison of two sequences and Gepard [15] and MUMmer [16] can be used for this purpose.
3. BUSCO
   BUSCO [13] is a collection of single copy orthologs that are conserved in 90% or more of the species sampled. BUSCO marker sets have been developed for insects, plants, etc. (*see* **Note 5**). These offer a useful metric for evaluation of the completeness of a genome assembly. BUSCO analysis can also be used to reveal the proportion of fragmented and missing markers that can be used as a proxy of the correctness of the assembly (*see* chapter on BUSCO for details).
4. Validation with a de novo transcriptome
   The gene space is one of the primary areas of interest in a new genome assembly, so it is important to evaluate the gene models for accuracy and completeness. A good quality de novo assembled transcriptome offers an independent source of validation of the structure of the genes annotated in the assembly. Moreover, this analysis can be done even before the assembly is fully annotated to assist in selecting candidate genome assemblies for further analysis. Please see chapter on MCOT for details on how to generate a de novo transcriptome assembly.

In this chapter, I have described a workflow to generate a high quality genome assembly from PacBio genomic data followed by evaluation of the assembly (*see* **Note 6**). The reader is referred to Richards et al. [2] for a review of other factors influencing arthropod genome sequencing. There are other useful tools that have not been mentioned here for the sake of brevity. For example, the Falcon Unzip assembler [26] used by Dudchenko et al. [6] is helpful for assembling heterozygous genomes as it is designed for assembling diploid organisms from PacBio reads. Optical maps created on the BioNano Genomics platform can be used for assembly validation, gap sizing, and very long range scaffolding of contigs. All the tools mentioned in this chapter have comprehensive user documentation, help forums, and tutorials on their websites (*see* Table 1) to guide the reader.

Although assemblies from long read sequencing contain fewer large structural errors in comparison to short read assemblies, careful validation is required nevertheless from multiple sources of evidence. It is our hope that the workflow described here provides sufficient background to aid in the design of an arthropod genome sequencing project, and the choice of technology and tools best suited for the purpose.

4 Notes

1. The input DNA is an important consideration for insect genomics as the amount of DNA that can be generated from a single individual for many species is in the range of a few hundred nanograms or even less. This amount is further reduced after cleanup and size selection to produce long fragments for library preparation. DNA extraction is typically performed on a pool of individuals to increase yield [2, 27]. However, this introduces additional complexity during genome assembly and may result in fragmented or chimeric assemblies. 10X, Oxford Nanopore, and Dovetail need DNA amounts ranging from a few to a few hundred nanograms. In comparison, PacBio requires at least a few micrograms of high molecular weight DNA before size selection. Long range scaffolding methods like Bionano, 10X, and Dovetail require DNA fragments in the range of 100 Kb to 1 Mb to be most effective. If multiple individuals are required to extract a sufficient amount of DNA, it is preferable to use individuals from a colony that has been inbred for many generations, as was done for *Cimex lectularius* [28]. Another choice is to use haploid individuals, if available. The *Bemisia tabaci* genome project used haploid males from a colony populated by progeny of a single female [29]. The quality of high molecular weight DNA should be checked (Qbit) to confirm that the majority of fragments are longer than 20 Kb.
2. Memory and disk space requirements will vary according to genome size and specific tool. Minimum RAM required for most assemblers is 500 Gb–1 Tb but it depends on the genome

size and complexity. The commands presume the computer has at 64 CPU cores.

3. Coverage of the genome is another important factor that determines quality of assembly. De novo assemblies require 50–80X coverage in most cases. However, this number depends on the repeat content of the genome. Assemblers like Canu can be optimized for a high coverage genome assembly. In order to deal with the heterozygosity of the input DNA sample, correction of all the raw reads (corOutCoverage=100) is recommended instead of the default (40X the genome size). An error rate of 0.013 (-utgOvlErrorRate=0.013) should be used for high coverage datasets. Please see the documentation for more information (https://canu.readthedocs.io/en/latest/faq.html#what-parameters-can-i-tweak).
4. Although PacBio and Oxford Nanopore sequences have long lengths that are usually constrained by limitations of DNA sample preparation, the primary challenges have historically been the high error rate per read and the low throughput. The high error rates have been offset by applying self-correction based on information from shorter reads for PacBio data sets. The final assembly should also be iteratively polished using Pilon with Illumina paired-end DNA sequences. RNA data can also be used to polish the gene space in the assembly but only SNPs and insertion or deletion errors should be corrected with this data.
5. We recommend the use of a clade specific marker set from BUSCO instead of the generic arthropod set as this enables more fine grained resolution of the completeness of the genome assembly [30]. A useful feature of a BUSCO analysis is the copy number of the marker genes. This is helpful in removal of the duplications that occur due to the presence of multiple individuals in the DNA sample, to create a haploid assembly. The location of the duplicated BUSCO markers can be used to filter out duplicated contigs from an assembly of multiple individuals. HaploMerger2 [31] can be used to build haploid subassemblies from a heterozygous diploid genome assembly. Jiao et al. [32] used evidence from Hi-C chromosome conformation capture to correct scaffolds generated from PacBio assemblies. This emphasizes the importance of validating an assembling from multiple and independent sources of evidence. If available, BACs and genetic maps should also be used to confirm the architecture of the scaffolds.
6. Many genome projects have opted to apply a combination of above mentioned tools and technologies to create the final assembly [33]. Most of the sequencing platforms are under rapid development resulting in frequent upgrades. Downstream software tools are also being actively developed in response. Eukaryotic assemblies have been reported from Oxford Nanopore data [34, 35] and this approach seems

promising for arthropod assemblies in the future. Therefore, the reader is encouraged to follow developments in the literature and on software tool websites (*see* Table 1).

Acknowledgments

I would like to thank Susan Brown and Michael Pfrender for the invitation to author this article. I would also like to thank my colleagues Prashant Hosmani and Mirella Flores for insightful discussions for troubleshooting assembly issues. This work was funded by USDA NIFA grant 2015-70016-23028 funded to Susan Brown and Lukas Mueller.

References

1. i5K Consortium (2013) The i5K initiative: advancing arthropod genomics for knowledge, human health, agriculture, and the environment. J Hered 104:595–600
2. Richards S, Murali SC (2015) Best practices in insect genome sequencing: what works and what doesn't. Curr Opin Insect Sci 7:1–7
3. T.I.A.G. Consortium (2010) Genome sequence of the pea aphid Acyrthosiphon pisum. PLoS Biol 8:e1000313
4. Maumus F, Fiston-Lavier A-S, Quesneville H (2015) Impact of transposable elements on insect genomes and biology. Curr Opin Insect Sci 7:30–36
5. Nene V, Wortman JR, Lawson D et al (2007) Genome sequence of aedes aegypti, a major arbovirus vector. Science 316:1718–1723
6. Dudchenko O, Batra SS, Omer AD et al (2017) De novo assembly of the Aedes aegypti genome using Hi–C yields chromosome-length scaffolds. Science
7. Koren S, Walenz BP, Berlin K et al (2017) Canu: scalable and accurate long-read assembly via adaptive k-mer weighting and repeat separation. Genome Res 27:722–736
8. Li H (2016) Minimap and miniasm: fast mapping and de novo assembly for noisy long sequences. Bioinformatics 32(14):2103–2110
9. English AC, Richards S, Han Y et al (2012) Mind the gap: upgrading genomes with pacific biosciences RS long-read sequencing technology. PLoS One 7:e47768
10. Yeo S, Coombe L, Chu J et al (2018) ARCS: assembly roundup by chromium scaffolding. Bioinformatics 34(5):725–731
11. Pryszcz LP, Gabaldón T (2016) Redundans: an assembly pipeline for highly heterozygous genomes. Nucleic Acids Res 44(12):e113
12. Walker BJ, Abeel T, Shea T et al (2014) Pilon: an integrated tool for comprehensive microbial variant detection and genome assembly improvement. PLoS One 9:e112963
13. Simao FA, Waterhouse RM, Ioannidis P et al (2015) BUSCO: assessing genome assembly and annotation completeness with single-copy orthologs. Bioinformatics 31:3210–3212
14. Wick RR, Schultz MB, Zobel J et al (2015) Bandage: interactive visualization of de novo genome assemblies. Bioinformatics 31:3350–3352
15. Krumsiek J, Arnold R, Rattei T (2007) Gepard: a rapid and sensitive tool for creating dotplots on genome scale. Bioinformatics 23:1026–1028
16. Kurtz S, Phillippy A, Delcher AL et al (2004) Versatile and open software for comparing large genomes. Genome Biol 5:R12
17. Li H (2018) Minimap2: fast pairwise alignment for long DNA sequences. Bioinformatics. https://doi.org/10.1093/bioinformatics/bty191
18. Adams MD, Celniker SE, Holt RA et al (2000) The genome sequence of drosophila melanogaster. Science 287(5461):2185–2195
19. Berlin K, Koren S, Chin C-S et al (2015) Assembling large genomes with single-molecule sequencing and locality-sensitive hashing. Nat Biotechnol 33(6):623–630
20. Miller JR, Delcher AL, Koren S et al (2008) Aggressive assembly of pyrosequencing reads with mates. Bioinformatics 24:2818–2824
21. Chaisson MJ, Tesler G, Ramaraj T et al (2012) Mapping single molecule sequencing reads using basic local alignment with successive refinement (BLASR): application and theory. BMC Bioinformatics 13:238
22. Putnam NH, O'Connell BL, Stites JC et al (2016) Chromosome-scale shotgun assembly

using an in vitro method for long-range linkage. Genome Res 26:342–350
23. Zheng GXY, Lau BT, Schnall-Levin M et al (2016) Haplotyping germline and cancer genomes with high-throughput linked-read sequencing. Nat Biotechnol 34:303–311
24. Rhoads A, Au KF (2015) PacBio sequencing and its applications. Genom Proteom Bioinf 13:278–289
25. Li H, Handsaker B, Wysoker A et al (2009) The sequence alignment/Map format and SAMtools. Bioinformatics 25:2078–2079
26. Chin C-S, Peluso P, Sedlazeck FJ et al (2016) Phased diploid genome assembly with single-molecule real-time sequencing. Nat Methods 13(12):1050–1054
27. Saha S, Hosmani PS, Villalobos-Ayala K et al (2017) Improved annotation of the insect vector of citrus greening disease: biocuration by a diverse genomics community, Database. bax032
28. Benoit JB, Adelman ZN, Reinhardt K et al (2016) Unique features of a global human ectoparasite identified through sequencing of the bed bug genome. Nat Commun 7:10165
29. Chen W, Hasegawa DK, Kaur N et al (2016) The draft genome of whitefly Bemisia tabaci MEAM1, a global crop pest, provides novel insights into virus transmission, host adaptation, and insecticide resistance. BMC Biol 14:110
30. Saha S, Hosmani P, Flores M, et al (2017) Using long reads, optical maps and long-range scaffolding to improve the Diaphorina citri genome
31. Huang S, Kang M, Xu A (2017) HaploMerger2: rebuilding both haploid sub-assemblies from high-heterozygosity diploid genome assembly. Bioinformatics 490:49–54
32. Jiao W-B, Garcia Accinelli G, Hartwig B et al (2017) Improving and correcting the contiguity of long-read genome assemblies of three plant species using optical mapping and chromosome conformation capture data. Genome Res 27(5):778–786
33. Mostovoy Y, Levy-Sakin M, Lam J et al (2016) A hybrid approach for de novo human genome sequence assembly and phasing. Nat Methods 13(7):587–590
34. Jain M, Koren S, Quick J et al (2018) Nanopore sequencing and assembly of a human genome with ultra-long reads. Nat Biotechnol 36:338–345
35. Erlich Y (2015) A vision for ubiquitous sequencing. Genome Res 25:1411–1416

Chapter 5

Integrated Modeling of Structural Genes Using MCuNovo

Xiaolong Cao and Haobo Jiang

Abstract

Correct modeling of protein-coding genes based on genome and cDNA data is a prerequisite for functional studies. Various programs such as MAKER, Cufflinks, Oases, and Trinity have been developed, each with advantages and drawbacks. Manual integration of different models for a single gene is cumbersome and becomes a daunting task for 14,000–18,000 genes in a typical holometabolous insect. We developed methods to evaluate the output of MAKER, Cufflinks, Oases and Trinity and select the best models to constitute the MCOT1.0 set for *Manduca sexta*, a biochemical model insect. To apply these methods in other organisms, we improved the algorithm (designated MCuNovo Gene Selector) and automated the data processing. In this chapter, we describe background information of algorithm development and how to prepare and run this program.

Key words Insect, Genomics, Transcriptome, Gene modeling, Python, Arthropod

1 Introduction

Next-generation sequencing (NGS) is widely used in genomics, transcriptomics, epigenetics, microRNA, and transcription factor research of animals, plants, and microbes [1–4]. While nearly a million insects account for over 75% of the total animal species on the planet, only 1000 have been sequenced to various depths for gene annotation, expression profiling, and functional exploration. *Drosophila melanogaster* and *Anopheles gambiae* are among the few that are extensively sequenced with genes manually annotated and regularly updated by researchers. For most insect genomes, genes are often modeled and annotated computationally, with a small portion manually curated. Low-quality models due to program limitations, a confusion of multiple models for a single gene, and a lack of data maintenance and ability to update gene models in most insect genome projects may haunt future functional studies in these species. As such, it is highly desirable that outputs of the gene prediction programs are compared to select the best model for each gene in existing and future arthropod genome projects.

Susan J. Brown and Michael E. Pfrender (eds.), *Insect Genomics: Methods and Protocols*, Methods in Molecular Biology, vol. 1858, https://doi.org/10.1007/978-1-4939-8775-7_5, © Springer Science+Business Media, LLC, part of Springer Nature 2019

1.1 Gene-Modeling Programs

Different programs and pipelines were developed for gene modeling and annotation [5]. Based on the data used, these programs can be divided into three categories: genome-based, genome- and transcriptome-based, and transcriptome-based. The MAKER pipeline [6], the TopHat-Cufflinks pipeline [7], and Trinity/Oases [8, 9] represent each of the three categories, respectively. Although these programs are commonly used in various genome projects, they all have advantages and limitations, as summarized previously [10] and in Fig. 1. Briefly, the MAKER pipeline combines ab initio gene predictions from SNAP [11], Augustus [12], and GeneMark-ES [13], and evidence-based gene predictions from EST and protein datasets to generate a GTF file that describes the gene structures in a genome. The MAKER pipeline is a good choice if RNA-seq data is not available, but splicing sites or alternative splicing forms may be inaccurate. The TopHat-Cufflinks pipeline works by aligning RNA-seq reads to the genome and predicting

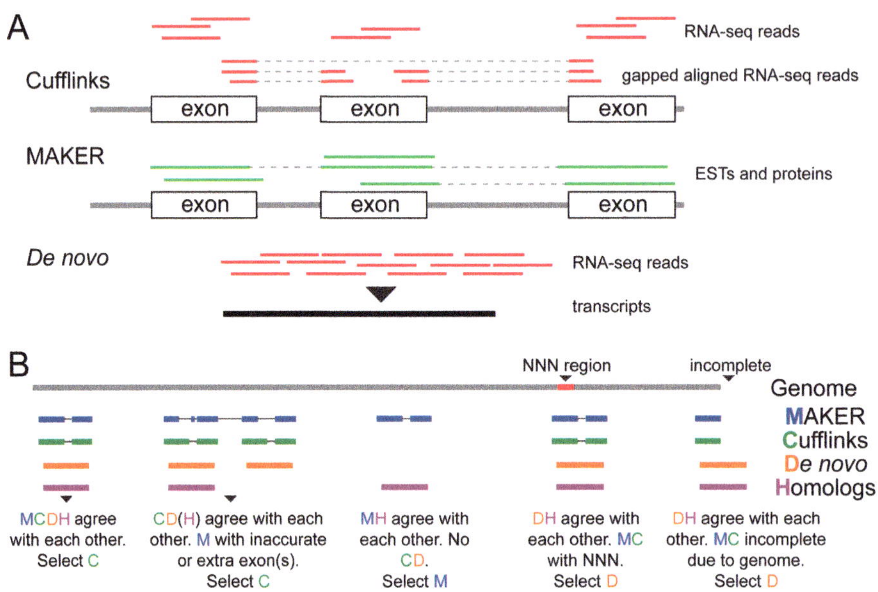

Fig. 1 Basic principles of MCuNovo and the gene-modeling programs behind MCuNovo. (**a**) Mechanism of MAKER, Cufflinks, MAKER, and de novo assemblers such as Trinity, Oases, and Bridger. MAKER gene annotation pipeline uses results from gene prediction programs including those using evidence from aligned cDNAs or proteins. Cufflinks works by aligning RNA-seq reads to the assembled genome with splicing sites supported by aligned reads with gaps. Trinity, Oases, and other de novo gene prediction programs directly assemble RNA-seq reads. (**b**) Selection of the best gene models from **M**AKER, **C**ufflinks, and de **novo** programs in various situations. MCuNovo tries to select best gene models from **M**AKER, **Cu**fflinks, and de **novo** assemblers. Cufflinks models are selected if all programs produce the same results. As Cufflinks works on the basis of experimental data, its models are favored over the corresponding MAKER models if they disagree. Gene transcripts not detected in the RNA-seq datasets (and, therefore, not in Cufflinks or de novo outputs) are be modeled by MAKER. When undetermined regions (NNNs instead of normal A/C/G/T bases) in a genome cause errors in MAKER and Cufflinks models, good models from the de novo assemblers are selected

accurate splicing sites from the gapped alignment of reads. However, it can only model genes that have sufficient read coverage, and may provide excessive alternative splice forms for some genes. As the outputs of MAKER and Cufflinks include coordinates of genes in the genome, some genes cannot be properly modeled if the corresponding genome regions are missing or problematic. De novo programs assemble RNA-seq reads independent of the genome, so errors in the genome sequence would not influence the modeled genes. Disadvantages of the de novo programs are lower sensitivity and higher redundancy of models than Cufflinks. The redundancy stems from sequencing errors, RNA editing, single nucleotide polymorphisms, and other issues.

Extensive RNA-seq data has become increasingly important in gene modeling. A single cDNA library only includes a small portion of all the structural genes in an organism, and most genes are differentially expressed in various tissues and life stages. However, with increased library diversity and sequencing depth, more genes can now be accurately modeled by Cufflinks and de novo programs. De novo gene models have become better sources for improving the accuracy of gene models from MAKER and Cufflinks. Pipelines such as PASA can take gene models from MAKER, Cufflinks, and de novo assemblers to generate unique maximal assemblies for manual improvements [14], but the outputs are still based on a genome that may have flaws including gaps, ambiguous bases, and misassembly errors. Many arthropod genomes suffer more or less from these problems, due to limited resources. On the other hand, since a complete set of accurate protein sequences is a prerequisite for functional research, it is highly desirable to have a program that compares the results from different gene-modeling programs in a less genome-dependent manner and selects the best models for all protein-coding genes in an organism.

1.2 Manual Improvement of Gene Models

Despite advances in modeling programs, manual gene annotation is still needed for the following reasons. First, modeling programs may provide inaccurate gene models. Some modeled transcripts are incomplete, or result from merged genes, or contain errors from predicted splice sites. Second, many genes are modeled with too many transcript isoforms, and many appear to have minor differences near the transcription start or end sites, but do not alter protein-coding sequences. Some gene models include incorrect boundaries of alternate exons, leading to early translational stops. Third, imperfections in the genome assembly cause problems, since most annotation work is based on genomes. Due to the limits of sequencing (read length, depth, and error), assembly programs, and complexity of the genome itself, most insect genomes are assembled into thousands of scaffolds. Each scaffold may contain many undetermined regions of various lengths [10], making it impossible to find the correct location of certain genes in the genome. Finally, for a genome project, only representative transcripts and proteins of genes are reported, to limit the total number of transcripts or proteins. In *D. melanogaster*, there are 304,788 transcripts, and many genes, each encoding more than 1000 transcripts and 100 proteins [15]. However, only 30,485 proteins or transcripts were reported in the official gene set (FlyBase, version 6.14, Feb. 9, 2017).

As structural genes are common subjects of biological research, manual annotation involves checking features of protein sequences translated from the modeled transcripts. Standards of correct gene models include: (1) the transcripts reside in the edges of genome scaffolds or regions with undetermined bases, (2) the translation products are complete at both N- and C-termini, (3) proteins from different modeling programs are identical or highly similar, and (4) the proteins have lengths and domain structures similar to their homologs in other species. De novo assemblies are useful references to verify gene models based on the genome, which may connect gaps between scaffolds and restore undetermined regions within gene models. In most cases, models with longer open reading frames are better and more similar to homologs in other species. Other information (e.g., signal peptide, domain structure, or published ESTs) also helps determine and select proper gene models. Still, in some cases, manual modification of exon boundaries and joining or splitting exons are required to obtain the correct gene models.

1.3 Development of MCuNovo

Manual validation and improvement of gene models is slow and tedious. Therefore, we developed an algorithm that follows principle of manual annotation, by comparing the results of MAKER, Cufflinks, Oases, and Trinity, and compiling the best gene models into a MCOT set. The principle is: for protein sequences derived from gene models predicted by these programs, select the one that is most similar in length to its homologs (Fig. 3). To do this, it is critical to find protein sequences predicted for the same gene by different gene-modeling programs and also identify homologs of this gene.

We applied this method to *Manduca sexta*, a biochemical model insect representing serious agricultural pests in the order of Lepidoptera [10]. In the work, we used the BLASTP program with a modified scoring matrix, in which scores for nonidentical amino acids were set to −100. As such, the resulted alignments between query and subject sequences are almost identical and adding up aligned length can be used to calculate the matched length (ML) between query and subject sequences. This modification is essential because, with normal scoring matrix, it is hard to calculate ML—some sequences from different genes can be very similar while sequences from the same gene might have lower identity due to gaps (Fig. 2). With proteins from MAKER or Cufflinks as

```
A  Identities = 203/224 (91%), Positives = 214/224 (96%), Gaps = 5/224 (2%)
   ...
   Query  119  ASLGEFPHMRAQGICDEKHLAAELQHVKQVLRDNKLQIPRRRHRNRVKPATVERVPVVLP  178
               A++G+    RA+GICDEKHLAAEL     +LRDNKLQIPRRRHR+RVKPATVERVPVVLP
   Sbjct  325  ATVGKSLFQRARGICDEKHLAAEL-----ILRDNKLQIPRRRHRDRVKPATVERVPVVLP  379

   Query  179  YVRGVTAKIGYALKRASIKTYFKPPKKIAQFLPPVKCHIPLQEPGVYKIDCNCGLSYIGQ  238
               YVRGVT KIGY+LKRASIKTYF PPKKI+QFLPPVKCHIPLQEPGVYKIDCNCGLSYIGQ
   Sbjct  380  YVRGVTDKIGYILKRASIKTYFEPPKKISQFLPPVKCHIPLQEPGVYKIDCNCGLSYIGQ  439
   ...

B  Identities = 235/275 (85%), Positives = 235/275 (85%), Gaps = 40/275 (15%)
   ...
   Query  49   EATWHERSGRRDMRRAPPAQTAVVALALSVLLRFVNAACYGRFHYTQHSSLTGSMGPGVP  108
               EATWHERSGRRDMRRAPPAQTAVVALALSVLLRFVNA               GSMGPGVP
   Sbjct  15   EATWHERSGRRDMRRAPPAQTAVVALALSVLLRFVNA---------------GSMGPGVP  59

   Query  109  ENITVTFLNPTTVRVSWSTTAELVEKYDVTYKPSDASYLRFVLPLHSSFVIENDEIDSLL  168
               ENITVTFLNPTTVRVSWSTTAELVEKYDVTYKPSDASY
   Sbjct  60   ENITVTFLNPTTVRVSWSTTAELVEKYDVTYKPSDASY----------------------  97
   ...
```

Fig. 2 Examples of a high identity alignment of sequences from different genes (**a**) and a low identity alignment of sequences from the same gene (**b**)

queries to search the Trinity and Oases assemblies, if the ML/QL (query length) was greater than 0.7, the Trinity or Oases models were considered to be the same transcripts as the query. Models for the same transcripts were selected based on their length and length of their homologs as previously described [10]. After removing highly similar sequences, we obtained MCOT1.0 for *M. sexta* with about 30,000 unique proteins, close to that of *D. melanogaster*.

To facilitate its application in other genome projects, we improved the original pipeline, and further developed the program. The original pipeline included many operations with Microsoft Excel and Python and manual checking for some genes, while those operations were now automated in the MCuNovo program, making it much faster and easier to use. We named the algorithm MCuNovo Gene Selector to indicate the fact that it is based on selection of the best gene models from MAKER (M), Cufflinks (Cu), and de novo (Novo) programs. We have applied MCuNovo to *Diaphorina citri Kuwayama* (Asian citrus psyllid), obtained about 30,000 unique proteins [16], and evaluated the gene set using BUSCO. BUSCO assesses completeness of gene sets with single-copy orthologs [17] and, with the profiles for arthropods, the BUSCO analysis showed that the MCOT 1.0 is the best in terms of completeness of orthologs for *M. sexta* and *D. citri* (Table 1). For insect species with genome sequences and RNA-seq data, running MCuNovo is a good starting point prior to manual annotation.

2 Materials

The MCuNovo program can be downloaded from https://github.com/ATPs/MCuNovo and run in Python 3 environment (https://www.continuum.io/downloads) with Biopython

Table 1
Assessment of the *M. sexta* and *D. citri* genome assemblies*

Species	Gene set	Transcript number	BUSCO notation assessment results
M. sexta	OGS 1.0	20,137	C:82% [D:10%], F:9.0%, M:8.4%, n:2675
	OGS 2.0	27,404	C:84% [D:37%], F:8.5%, M:7.1%, n:2675
	Cufflinks 3.0	53,102	C:76% [D:47%], F:8.7%, M:14%, n:2675
	Trinity 4.0	155,825	C:90% [D:51%], F:4.6%, M:4.3%, n:2675
	Oases 4.0	304,367	C:78% [D:72%], F:13%, M:7.9%, n:2675
	Trinity W	43,871	C:54% [D:22%], F:21%, M:24%, n:2675
	MCOT 1.0	31,166	C:93% [D:32%], F:3.4%, M:2.5%, n:2675
D. citri	MAKER	15,322	C:46% [D:7.9%], F:25%, M:28%, n:2675
	MCOT 1.0	30,562	C:90% [D:34%], F:4.7%, M:4.6%, n:2675

*BUSCO version 1.1 (Simão et al., 2015) was run to assess qualities of the *M. sexta* and *D. citri* gene sets against 2675 benchmarking universal single-copy orthologs in arthropods. The BUSCO notations are: *C* for complete [*D* for duplicated], *F* for fragmented, *M* for missing, and *n* for gene number

(http://biopython.org/DIST/docs/install/Installation.html). If Anaconda is installed as the python environment in a Windows system, Biopython can be installed in Command Prompt using the command line: "conda install biopython." BLAST+ software can be downloaded from ftp://ftp.ncbi.nlm.nih.gov/blast/executables/blast+/LATEST/ [18].

MCuNovo, designed to select the best gene models generated by different programs, needs four fasta files of protein/transcript sequences, one from MAKER or a similar program, one from Cufflinks, and two from de novo assemblers such as Trinity [8], Oases [9] and Bridger [19]. Transcript sequences can be translated to proteins using TransDecoder (http://transdecoder.github.io/) or an equivalent program. BLAST+ and TransDecoder are run in a Linux environment. MCuNovo also uses protein sequences from other species to improve the gene models and, for insects, arthropod sequences (ftp://ftp.ebi.ac.uk/pub/databases/fastafiles/uniprot/uniprotkb_arthropoda.gz) should be downloaded for use as reference.

3 Methods

3.1 Preparation of Protein Sequences from Different Programs

For a specific species, genes are modeled in the form of transcripts according to the manuals of MAKER, Cufflinks, and the de novo assemblers (*see* **Note 1**). Transcripts are translated to proteins using TransDecoder in a Linux environment (*see* **Note 2**). The typical scripts are:

```
TransDecoder.LongOrfs -t target_transcripts.fasta
TransDecoder.Predict -t target_transcripts.fasta
```

After this step, four files of protein sequences are ready for comparison.

3.1.1 Removal of Duplicated Protein Sequences (Optional)

While MCuNovo generates a file of nonredundant protein sequences, to save time, we suggest reducing redundancy before running by the following steps. We provide the "fastaFileRemoveDup.py" module, which yields a new file without duplicated sequences. By default, the allowed mismatching rate is set to 0.02 to determine whether or not two sequences are duplicates.

```
python fastaFileRemoveDup.py -h
python fastaFileRemoveDup.py trinity.protein
```

In our experience, this optional step considerably reduced the number of protein sequences in Cufflinks, Trinity, and Oases, and the newly generated files are used in the following steps.

3.2 Preparation of Reference Sequences

Reference sequences can be Arthropod proteins, which can be downloaded as described above, or other homologous sequences. Arthropod sequences from UniProt, which are located in the file "uniprotkb_arthropoda," are used as an example to better explain the process.

3.2.1 Removal of Low-Quality Sequences (Optional)

To remove incomplete (as indicated by a key word "fragment" in a protein name) or imperfect (with "X" in a protein sequence) sequences, open the "fastaFileClean.py" file with a text editor, change the string behind "filename =" to full path of the reference fasta file, and then run the file. The full path should look like "F:\crInsects\cruniprotkb_arthropoda20170117" in Windows as shown in the file, or "/Insects/uniprotkb_arthropoda20170117" in Linux. To run this file, the script in the command line is "python PATH_TO_FOLDER/fastaFileClean.py," where PATH_TO_-FOLDER is the folder of MCuNovo program. Or in Windows, after installing Anaconda Python, open the "fastaFileClean.py" file in spyder, and run the file by clicking the "run file" button or by copying and pasting the content to the IPython console. The parameters can be adjusted to meet different needs.

3.3 Preparation of BLAST+ Program

We suggest that users install and run BLAST+ program in the Linux environment, and provide two ways to prepare the program. The methods can be adjusted for other operating systems.

3.3.1 Method 1: Pre-built Binary Version

Download and decompress the binary version of BLAST+ 2.6.0 for Linux. Replace the "blastp" file in the 'ncbi-blast-2.6.0+/bin/" with the "blastp" file in MCuNovo folder, and run the "chmod" command line below to gain the all permissions for the pre-built "blastp" program for MCuNovo. New BLOSUM45 in the modified "blastp" is BLOSUM62 with matching scores of nonidentical amino acids all changed to −100. (The original BLOSUM45 is deleted whereas BLOSUM62 is unchanged). The details on how the scoring matrix is changed can be found in the "tables" folder inside MCuNovo.

```
cd WORK_PATH
wget  ftp://ftp.ncbi.nlm.nih.gov/blast/executables/blast+/
2.6.0/ncbi-blast-2.6.0+-x64-linux.tar.gz
tar xzf ncbi-blast-2.6.0+-x64-linux.tar.gz
chmod 777 WORK_PATH/ncbi-blast-2.2.31+/bin/blastp
```

3.3.2 Method 2: A Version Built from Source Code

Download and decompress the source code of the newest version of BLAST+ (2.6.0). Replace the folder "ncbi-blast-2.6.0+-src/c++/src/util/tables" with the "tables" folder in MCuNovo. Build the program according to the manual (https://www.ncbi.nlm.nih.gov/books/NBK279671/). The compiled executables are in "ncbi-blast-2.6.0+-src/c++/ReleaseMT/bin." In the Linux system tested, we successfully built the program with gcc-4.9.2 using the "--without-boost" option during program building as shown below.

```
cd WORK_PATH
wget  ftp://ftp.ncbi.nlm.nih.gov/blast/executables/blast+/LAT-
EST/ncbi-blast-2.6.0+-src.tar.gz
tar xzf ncbi-blast-2.6.0+-src.tar.gz
cd ./ncbi-blast-2.6.0+-src/c++
./configure --without-boost
cd ReleaseMT/build
make all_r
```

3.4 Protein Sequence Comparison

Currently, MCuNovo uses the output of BLASTP in hit-table format to perform sequence comparison. The modified BLASTP program we built previously is used in this step.

3.4.1 Building a Protein Database for BLASTP Search

FASTA files from the species under study, named as "maker.fasta," "cufflinks.fasta," "oases.fasta," and "trinity.fasta," include proteins modeled by the programs in the file names (*e.g.*, MAKER, Cufflinks). The file "uniprot.fasta" includes current reference protein sequences from arthropods for homolog comparison and improvement. These files are used to construct a sequence database according to the manual of BLAST+ program (https://www.ncbi.nlm.nih.gov/books/NBK279690/). If the files are prepared in Windows or MacOS, the "dos2unix" function should be used to make them compatible with Linux. Example codes are shown here.

```
dos2unix cufflinks.fasta
segmasker  -in  cufflinks.fasta  -infmt  fasta  -parse_seqids
-outfmt maskinfo_asn1_bin -out cufflinks.asnb
makeblastdb -in cufflinks.fasta -input_type fasta -dbtype prot
-parse_seqids -mask_data cufflinks.asnb -out cufflinksDB -ti-
tle "blastp database cufflinks"
```

3.4.2 Sequence Comparisons of Proteins from Different Gene-Modeling Programs

The number of identical residues between two protein sequences is used by MCuNovo to select proteins from the gene-modeling programs. To ensure BLASTP only outputs nearly perfect alignments, we used the BLASTP built previously, using the modified BLOSUM45 matrix with the gap-opening and gap-extending penalty set to maximum. MCuNovo uses the hit-table ("-outfmt 6") from BLASTP. In total, there are ten comparisons, M2C, C2M, M2O, O2M, M2T, T2M, C2O, O2C, C2T, and T2C, where M, C, O, and T represent proteins from MAKER, Cufflinks, Oases, and Trinity, respectively.

```
blastp -db cufflinksDB -comp_based_stats F -query maker.fasta
-gapopen 32767 -gapextend 32767 -out M2C.txt -matrix BLOSUM45
-num_threads 12 -outfmt 6 -max_target_seqs 20
* set -num_threads based on the number of cores in CPU
```

3.4.3 Sequence Comparison of Modeled Proteins with Arthropod Reference Proteins

Similarities with reference sequences are used by MCuNovo to select proteins from different programs. Four BLASTP comparisons are performed, including M2U, C2U, O2U, and T2U, where U represents reference sequences (*e.g.*, arthropod proteins in the UniProt database).

```
blastp -db AthropodaDB -query maker.fasta -out M2U.txt -outfmt
6 -num_threads 12 -max_target_seqs 20 -matrix BLOSUM62
```

3.5 Preparation of a Protein-to-Gene File (Optional)

Some gene transcripts are alternatively spliced and, hence, translated into several proteins. MCuNovo can assign the same gene ID for protein variants encoded by that gene, by taking an input file with lines of protein name and gene name separated by a tab. Due to different naming strategies of the gene-modeling programs, it is difficult to write a program that works universally. The "MCuNovoGenerateProtein2Gene.py" module is compatible with most gene-modeling programs. Codes below show how to run this module or get help. This is not required for all protein–gene pairs, and Cufflinks protein–gene pairs are good enough in most cases.

```
python MCuNovoGenerateProtein2Gene.py -h
python MCuNovoGenerateProtein2Gene.py --cufflinks cufflinks.
fasta cufflinks.protein --denovo trinity.protein oases.protein
--maker maker.protein --out protein2gene.txt
```

3.6 Running MCuNovo

The basic working scheme of MCuNovo [10] is briefly shown in Fig. 3. "MCuNovoRunTest.py" file is used to test the main MCuNovo. In the Command Prompt of Windows or terminal of Linux, run the file with the scripts "python PATH_TO_ FOLDER/MCuNovoRunTest.py," where PATH_TO_FOLDER is a folder of MCuNovo program. Alternatively, open the "MCuNovoRunTest.py"

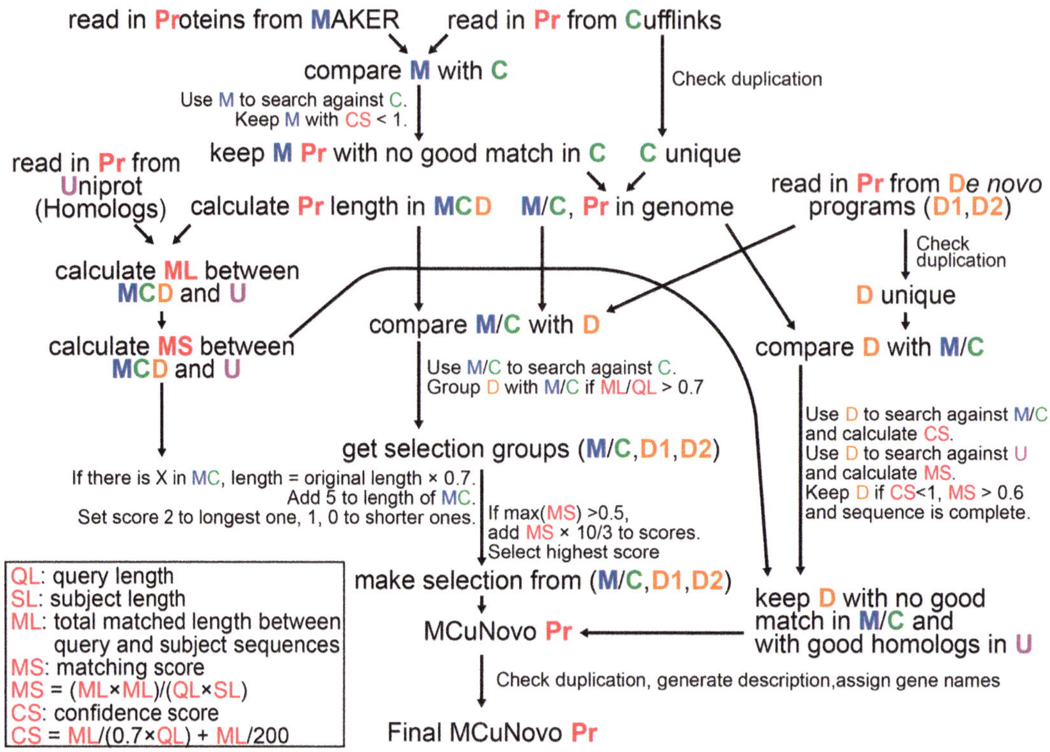

Fig. 3 Scheme of MCuNovo Gene Selector. M, C, D1/D2, and U represent protein sequences modeled by MAKER, Cufflinks, De novo programs, and proteins from Uniprot, respectively. Briefly, M is compared with C, and sequences with CS $<$ 1 were combined with unique sequences in C. The combined M/C sequences are compared with D1 and D2, and D1/D2 sequences will be grouped with M/C if ML/QL $>$ 0.7 to form selection groups. A decision will be made to select the best one by evaluate the length of protein sequences in the same selection group, the existence of undetermined residues (amino acid X), and the length of homologous sequences (shown in the figure). The selected protein is then combined with good quality proteins from D to get the final MCuNovo results. Sequences in D will be kept if they had no good match in M/C (CS $<$ 1) and had good homologs in U (MS $>$ 0.6). CS stands for confidence score. If CS $<$ 1, we may consider that the query sequence does not have good match in the subject sequences. MS stands for matching score, and represents the similarity between the query and subject sequences. CS $=$ ML/(0.7 $\times$ QL) + ML/200. The "coverage" and "min_length" are 0.7 and 200 in "MCuNovoRun.py." The threshold values used in the figure are empirical values which worked well in our experiences

file in spyder and run it by clicking the "run file" button. A "MCDpeptides.txt" file will be generated in the test folder, and the name line for proteins in "MCDpeptides.txt" includes: MCuNovo (gene name, length), original protein before selection (name, length), selected protein (name, length), MAKER with the best matching score for the selected protein (name, length, matched length), Cufflinks (name, length, matched length), de novo program 1 (name, length, matched length), de novo program 2 (name, length, matched length), and reference (name, length, matched length, description), as shown in Fig. 4.

MCD0000003.1 Len:427
MCuNovo (gene name, length)
AGAP028669-PA 427 AGAP028669-PA 427
original protein before selection (name, length), selected protein (name, length),
AGAP028669-PA 427 427 NA 0 0 NA 0 0 NA 0 0
MAKER (name, length, ML), Cufflinks (name, length, ML), de novo 1 (name, length, ML), de novo 2 (name, length, ML)
NA 0 0 NA
reference (name, length, ML, description)

MCD0000003.1
gene ID / protein ID
protein #, 0 if only one protein in that gene

MCD0000003.2 Len:413
MCuNovo (gene name, length)
TCONS_00037670|m.634266 408 3L168526i2|m.691824 413
original protein before selection (name, length), selected protein (name, length),
AGAP028669-PB 290 227 TCONS_00037670|m.634266 408 312 3L168526i2|m.691824 413 413 NA 0 0
MAKER (name, length, ML), Cufflinks (name, length, ML), de novo 1 (name, length, ML), de novo 2 (name, length, ML)
TR:A0A0E4G8E9_ANOGA 342 318 Uncharacterized protein Anopheles gambiae
reference (name, length, ML, description)

Fig. 4 An example for MCuNovo gene naming. The name line contains 24 elements, as described in Subheading 3.6. *ML* matched length. *NA* data not available

Because MCuNovo takes many input files, the "MCuNovoRun.py" file is provided and users need to modify the content based on paths where the input files are located (*see* **Note 3**). The input files include protein models from four different modeling programs, one reference protein file, ten files for comparisons between the modeling programs, four files for comparisons between the four model files and the reference file, and one protein-to-gene file (optional). The details about each parameter are described in the "MCuNovoRun.py" file. After proper modifications, the file can be run in the same way as for "MCuNovoRunTest.py." For users who are familiar with Python, content of the "MCuNovoGeneSelectorPep" function in the "MCuNovoGeneSelectorMain.py" can be run in blocks in the IPython console to monitor the processing of intermediate data.

3.7 Acquisition of DNA Sequences

DNA sequence acquisition, a function not included in MCuNovo, can be done by using names of the selected proteins in the output file of MCuNovo to retrieve the corresponding DNA sequences.

4 Notes

1. New gene-modeling programs are developed, including the BRAKER pipeline [20], the HiSTAT2-StringTie pipeline [21], and Bridger/BinPacker [19, 22] serving as alternatives of MAKER, Cufflinks, and Trinity/Oases, respectively. Their results could be treated in the same way.

2. If a file is transferred from Windows to Linux, and the program runs abnormally, the "dos2unix" function should be used to convert the file for the Linux system.
3. The "coverage" and "min_length" values in "MCuNovoRun.py" should not be changed if users are unclear about the use of these values. These parameters define the confidence score (CS) (Fig. 3). "coverage" is also used to determine whether M/C and D should be grouped together.

Acknowledgments

This study is supported by NIH grants GM58634 and AI112662. This work was approved for publication by the Director of Oklahoma Agricultural Experimental Station and supported in part under project OKLO2450.

References

1. Metzker ML (2010) Sequencing technologies—the next generation. Nat Rev Genet 11(1):31–46
2. Koboldt DC et al (2013) The next-generation sequencing revolution and its impact on genomics. Cell 155(1):27–38
3. Wang Z, Gerstein M, Snyder M (2009) RNA-Seq: a revolutionary tool for transcriptomics. Nat Rev Genet 10(1):57–63
4. Park PJ (2009) ChIP–seq: advantages and challenges of a maturing technology. Nat Rev Genet 10(10):669–680
5. Yandell M, Ence D (2012) A beginner's guide to eukaryotic genome annotation. Nat Rev Genet 13(5):329–342
6. Holt C, Yandell M (2011) MAKER2: an annotation pipeline and genome-database management tool for second-generation genome projects. BMC Bioinformatics 12:491
7. Trapnell C et al (2012) Differential gene and transcript expression analysis of RNA-seq experiments with TopHat and Cufflinks. Nat Protoc 7(3):562–578
8. Grabherr M et al (2011) Full-length transcriptome assembly from RNA-Seq data without a reference genome. Nat Biotechnol 29(7):644–652
9. Schulz M et al (2012) Oases: robust de novo RNA-seq assembly across the dynamic range of expression levels. Bioinformatics (Oxford, England) 28(8):1086–1092
10. Cao X, Jiang H (2015) Integrated modeling of protein-coding genes in the Manduca sexta genome using RNA-Seq data from the biochemical model insect. Insect Biochem Mol Biol 62:2–10
11. Korf I (2004) Gene finding in novel genomes. BMC Bioinformatics 5:59
12. Stanke M, Waack S (2003) Gene prediction with a hidden Markov model and a new intron submodel. Bioinformatics 19(Suppl 2):ii215–ii225
13. Lomsadze A et al (2005) Gene identification in novel eukaryotic genomes by self-training algorithm. Nucleic Acids Res 33(20):6494–6506
14. Haas BJ et al (2008) Automated eukaryotic gene structure annotation using EVidenceModeler and the Program to Assemble Spliced Alignments. Genome Biol 9(1):1–22
15. Brown JB et al (2014) Diversity and dynamics of the Drosophila transcriptome. Nature 512(7515):393–399
16. Saha S et al (2017) Improved annotation of the insect vector of citrus greening disease: Biocuration by a diverse genomics community. Database 1–20
17. Simão FA, Waterhouse RM, Ioannidis P, Kriventseva EV, Zdobnov EM (2015) BUSCO: assessing genome assembly and annotation completeness with single-copy orthologs. Bioinformatics 31(19):3210–3212
18. Camacho C et al (2009) BLAST+: architecture and applications. BMC Bioinformatics 10:421
19. Chang Z et al (2015) Bridger: a new framework for de novo transcriptome assembly using RNA-seq data. Genome Biol 16:30

20. Hoff KJ et al (2016) BRAKER1: unsupervised RNA-Seq-based genome annotation with GeneMark-ET and AUGUSTUS. Bioinformatics 32(5):767–769
21. Pertea M et al (2016) Transcript-level expression analysis of RNA-seq experiments with HISAT, StringTie and Ballgown. Nat Protoc 11(9):1650–1667
22. Liu J et al (2016) BinPacker: packing-based De Novo transcriptome assembly from RNA-seq data. PLoS Comput Biol 12(2): e1004772

Chapter 6

Using BUSCO to Assess Insect Genomic Resources

Robert M. Waterhouse, Mathieu Seppey, Felipe A. Simão, and Evgeny M. Zdobnov

Abstract

The increasing affordability of sequencing technologies offers many new and exciting opportunities to address a diverse array of biological questions. This is evidenced in entomological research by numerous genomics and transcriptomics studies that attempt to decipher the often complex relationships among different species or orders and to build "omics" resources to drive advancement of the molecular understanding of insect biology. Being able to gauge the quality of the sequencing data is of critical importance to understanding the potential limitations on the types of questions that these data can be reliably used to address. This chapter details the use of the Benchmarking Universal Single-Copy Orthologue (BUSCO) assessment tool to estimate the completeness of transcriptomes, genome assemblies, and annotated gene sets in terms of their expected gene content.

Key words Genomics, Transcriptomics, Genome annotation, Completeness assessment, Single-copy orthologues

1 Introduction

Advances in genomics technologies mean that high-throughput nucleotide sequencing has become a relatively low-cost and thus widely accessible tool with numerous applications in biological research. Nevertheless, as researchers in the field know only too well, technical issues, e.g., sample preparation, as well as biological complexities, e.g., large genome sizes, can present substantial challenges to successfully building high-quality genomics resources [1]. Most leading technologies offer in-house sequencing accuracy estimates, and several computational tools allow for detailed assessments of the performance of sequencing and assembly strategies, e.g., `QUAST` [2] and `REAPR` [3] genome assembly quality evaluators. Metrics such as contig or scaffold N50 values (half of the total assembly span is made up of contigs or scaffolds of length N50 or longer) offer a summary-statistic view of genome assembly contiguity. Scaffold counts and N50 values from a representative

Susan J. Brown and Michael E. Pfrender (eds.), *Insect Genomics: Methods and Protocols*, Methods in Molecular Biology, vol. 1858, https://doi.org/10.1007/978-1-4939-8775-7_6,

Table 1
Assembly statistics and BUSCO assessment results from a representative selection of recently published draft insect genomes

Organism	Species	Assembly size (Mbps)	Number of scaffolds	Scaffold N50 (Kbps)	BUSCO completeness	Publication
Fruit fly	*Drosophila serrata*	198.3	1360	942.6	C:94.1% F:2.5%, M:1.3%	Allen et al. 2017 [4]
Postman butterfly	*Heliconius melpomene*	275.2	795	2103	C:81.6% F:11.1%, M:7.3%	Davey et al. 2016 [5]
Tobacco hornworm moth	*Manduca sexta*	419.4	20,871	664	C:86.4% F:8.4%, M:5.2%	Kanost et al. 2016 [6]
Mycalesine butterfly	*Bicyclus anynana*	475.4	10,800	638	C:98.3% F:0.9%, M:0.8%	Nowell et al. 2017 [7]
Mediterranean fruit fly	*Ceratitis capitata*	479	1806	4118	C:95.6% F:3.4%, M:1.0%	Papanicolaou et al. 2016 [8]
Bed bug	*Cimex lectularius*	650.5	1402	7173	C:78.6% F:14.0%, M:7.4%	Benoit et al. 2016 [9]
Asian longhorned beetle	*Anoplophora glabripennis*	710	10,473	659	C:85.7% F:11.0%, M:3.3%	McKenna et al. 2016 [10]
Banded demoiselle	*Calopteryx splendens*	1630	8896	422	C:53.5% F:31.8%, M:14.7%	Ioannidis et al. 2017 [11]

BUSCO completeness: *C* complete, *F* fragmented, *M* missing. Species are ordered from the smallest to the largest assembly size and all reported values were retrieved directly from each of the publications listed

selection of recently published draft insect genomes show that some are currently rather fragmented and will require considerable improvement efforts to reach near-chromosomal-level status (Table 1). However, as a major goal of many genomics studies is to catalogue the complete repertoire of protein-coding genes to facilitate subsequent detailed molecular biology experiments, it is important to also assess the quality of these resources with respect to their completeness in terms of their expected gene content.

The Benchmarking Universal Single-Copy Orthologue (BUSCO) assessment tool [12, 13] implements such quantifications of completeness for assembled genomes and transcriptomes, as well as annotated protein-coding gene sets. The assessment tool identifies matches to sets of genes that are expected to be present as single-copy orthologues in a given taxonomic group. This expectation is defined by surveying major species clades with numerous sequenced and annotated genomes to identify near-universally-present single-copy orthologues, using the OrthoDB [14] catalogue of orthologues (http://www.orthodb.org). For arthropods, BUSCO currently provides five assessment lineages: Arthropoda, Insecta, Endopterygota, Hymenoptera, and Diptera (http://busco.ezlab.org). The evolutionary filter for genes that are almost always present as single-copy orthologues across a given clade, i.e., genes evolving under "single-copy control" [15, 16], means that they are expected to be present in any newly sequenced species from the same taxonomic group. Quantifying proportions of BUSCOs that can be reliably identified from different genomic resources therefore provides like-for-like estimates of their relative completeness that complement other quality metrics. Importantly, this means that even if a draft genome assembly is still rather fragmented, good BUSCO completeness results allow researchers to proceed with confidence knowing that they have managed to capture most of the expected protein-coding gene repertoire. The examples in Table 1 illustrate how scaffold counts or N50 values are not necessarily predictive of BUSCO completeness, highlighting the importance of such assessments to ensure transparent and intuitive genomic resource quality measures for the benefit of the entire research community.

This chapter presents step-by-step examples of using BUSCO to assess the completeness of different insect genomics resources, with sufficient detail to allow even those unfamiliar with command line computing to run their own assessments. The assessment process consists of running a computational pipeline to identify and then classify BUSCO matches from genome assemblies, annotated gene sets, or transcriptomes, using HMMER [17] hidden Markov models (HMMs). For transcriptomes the longest open reading frames are assessed, while for genome assessments, gene models are first built using ab initio gene prediction with Augustus [18] for the potential matches identified using tBLASTn [19] searches. Matches that meet the BUSCO HMM score cutoffs are classified as *"complete"* if their lengths fall within BUSCO profile length expectations, and if found more than once they are classified as *"duplicated"*. Those that do not meet the length requirements are considered as partial matches and are classified as *"fragmented"*, and BUSCOs without matches that pass the thresholds are classified as *"missing"*. In this way, the assessments provide an intuitive quantification of the completeness of different genomics datasets in terms of expected gene content.

2 Materials

Before running BUSCO assessments, users are required to first set up the BUSCO software and its dependencies on their computer system and make sure that the data they wish to analyze adhere to the correct formats. These are outlined below, and users are encouraged to visit the website and read the user guide for further detailed information (http://busco.ezlab.org).

2.1 Software Setup

1. `BUSCO` has been developed in Python and tested on Linux operating systems and it is therefore recommended to use a Linux machine for running `BUSCO` and its dependencies.
2. The software distribution is available from a public GitLab project where it can be downloaded or preferably (*see* **Note 1**) cloned using a git client:

```
$ git clone https://gitlab.com/ezlab/busco.git
```

3. As well as Python, the following software packages are `BUSCO` dependencies and thus must also be installed on the system:

```
HMMER (v3.1b2) from http://hmmer.org
NCBI BLAST+ from https://ftp.ncbi.nlm.nih.gov/blast/executables/blast+ (see Note 2)
Augustus (v3.2.1 or above) from http://bioinf.uni-greifswald.de/augustus (see Note 3)
```

4. BUSCO `v3` is installed on the system by executing the `setup.py` script (*see* **Note 4**):

```
$ sudo python setup.py install (with root privileges).
$ python setup.py install --user (with only user privileges)
```

5. `BUSCO v3` setup is controlled with a user-editable configuration file. The `config.ini.default` file in the `BUSCO` "`config`" directory must first be copied to `config.ini` before editing. In this copied file, users must declare the paths to all dependencies (this simply tells `BUSCO` where they are installed on your system). Users may also use this `config.ini` file to define the input parameters for a particular analysis, but should be aware that providing input parameters through the command line will override those defined in the `config.ini` file.
6. Users without access to a Linux machine or cluster may instead use the `BUSCO` virtual machine (VM). The `BUSCO` VM was built using OSboxes (http://www.osboxes.org), it comes with the `BUSCO` software and its dependencies already preinstalled and can be downloaded from http://busco.ezlab.org. To run the VM, users need to first download and install a VM manager that is compatible with their system (e.g., Windows, Linux, Macintosh, or Solaris) such as VMware (https://www.vmware.com) or VirtualBox (https://www.virtualbox.org).

7. It is highly recommended to first run a test using the sample data provided as part of the BUSCO software distribution. Execute the following commands and compare the final output "run_TEST" with the provided files in "sample_data/run_SAMPLE".

```
$ python scripts/run_BUSCO.py --in sample_data/target.fa
--out TEST
--lineage_path sample_data/example --mode genome
```

2.2 Input Data

1. This chapter details the assessment of several publicly available mosquito genome assemblies and their annotated gene sets from genome sequencing projects [20–22] that can all be downloaded from VectorBase [23] (https://www.vectorbase.org/downloads).
2. The examples used for running transcriptome assessments were selected from publicly available hymenopteran datasets generated as part of large-scale insect transcriptomics studies [24, 25] that can all be downloaded from the NCBI's Transcriptome Shotgun Assembly (TSA) Sequence Database (https://www.ncbi.nlm.nih.gov/genbank/tsa).
3. Input sequence data for genome, transcriptome, or gene set assessments should be provided in standard FASTA format. Files that contain nonstandard nucleotides or amino acids in the sequence lines, or non-alphanumeric or non-ASCII characters in the header lines, could cause errors and therefore these should be avoided wherever possible.
4. Pre-processing of the input data is required in order to obtain true estimates of the numbers of duplicated BUSCOs for annotated gene sets and transcriptomes, these should be pre-processed to select just one representative transcript per gene (*see* **Note 5**).
5. The lineage datasets used for BUSCO assessments are not provided with the software distribution. Instead, users should download the appropriate lineage dataset(s) from http://busco.ezlab.org (*see* **Note 6**). For example, this chapter uses genomic data from dipterans and hymenopterans, so:

```
$ wget http://busco.ezlab.org/datasets/diptera_odb9.tar.gz
$ wget http://busco.ezlab.org/datasets/hymenoptera_odb9.
tar.gz
```

6. Each downloaded lineage dataset will need to be unpacked and decompressed before it can be used, for example:

```
$ tar -xf diptera_odb9.tar.gz
$ tar -xf hymenoptera_odb9.tar.gz
```

3 Methods

With the BUSCO software and its dependencies correctly set up, the relevant BUSCO lineage datasets downloaded and unpacked, and some example insect genomics datasets downloaded and pre-processed (if required), genome, gene set, and transcriptome assessments can now be performed.

3.1 Genome Assessments

1. To run genome mode assessments BUSCO needs to know the location of the AUGUSTUS configuration directory, so the "config" path must first be declared as an environment variable (*see* **Note 3**):

```
$ export AUGUSTUS_CONFIG_PATH="/path/to/AUGUSTUS/augus-
tus-3.2.3/config/"
```

2. The command to launch a genome assembly assessment is made up of four mandatory argument-value pairs that follow the python call to the run_BUSCO.py script:

```
--in SEQUENCE_FILE          path to your FASTA file, here your genome
--out NAME                  a short name that identifies your analysis run
--lineage_path LINEAGE      path to the BUSCO lineage dataset directory
--mode MODE                 specify which analysis mode to run, here "genome"
```

So to launch an assessment of the genome assembly of the *Anopheles arabiensis* mosquito using the dipteran lineage dataset the command would be (*see* **Note 7**):

```
$ python /path/to/busco/scripts/run_BUSCO.py
--in /path/to/mosquito/genomes/Anopheles-arabiensis-D1-genome.fs
--out AARAD1
--lineage /path/to/lineage/dataset/diptera_odb9
--mode genome
```

3. There are several additional argument-value pairs that are optional and allow users to change the default values of various settings, e.g., if the user's system has access to multiple computing cores, then they can take advantage of this using the --cpu argument (CPU, central processing unit), or the e-value cutoff for TBLASTN searches can be changed with the --evalue argument (*see* **Note 8**).
4. One of the most important optional arguments to consider for genome assembly assessments is the choice of AUGUSTUS pre-trained species-specific gene prediction parameters. Each BUSCO lineage dataset has a predefined default selection, e.g.,

for the Diptera lineage the default is "`fly`", which are Augustus gene prediction parameters pretrained on the fruit fly, `Drosophila melanogaster` (*see* **Note 9**).

5. Running the above assessment of the 246.6 megabasepair (Mbp) *Anopheles arabiensis* genome assembly using the dipteran lineage dataset on 12 CPUs with otherwise default options should take approximately 4 h (*see* **Note 10**).

3.2 Gene Set Assessments

1. Launching an assessment of an annotated gene set follows the same basic rules as for genome assemblies, with the same four mandatory argument-value pairs that follow the python call to the `run_BUSCO.py` script:

```
--in SEQUENCE_FILE       path to your FASTA file, here your proteins
--out NAME               a short name that identifies your analysis run
--lineage_path LINEAGE   path to the BUSCO lineage dataset directory
--mode MODE              specify which analysis mode to run, here "proteins"
```

So to launch an assessment of the *Anopheles arabiensis* annotated gene set (version AaraD1.6) using the dipteran lineage dataset the command would be:

```
$ python /path/to/busco/scripts/run_BUSCO.py
--in /path/to/mosquito/genesets/Anopheles-arabiensis-D1-proteins-1.6.fs
--out AARAD16
--lineage /path/to/lineage/dataset/diptera_odb9
--mode proteins
```

2. Running this assessment of the 13,452 *Anopheles arabiensis* (AaraD1.6) protein-coding genes using the dipteran lineage dataset on 4 CPUs with otherwise default options should take approximately 40 min (*see* **Notes 10** and **11**).

3.3 Transcriptome Assessments

1. Transcriptome assessments are launched with the same four mandatory argument-value pairs that follow the python call to the `run_BUSCO.py` script:

```
--in SEQUENCE_FILE       path to your FASTA file, here your transcripts
--out NAME               a short name that identifies your analysis run
--lineage_path LINEAGE   path to the BUSCO lineage dataset directory
--mode MODE              specify which analysis mode to run, here "transcriptome"
```

So to launch an assessment of the transcriptome from an adult `Pelecinus polyturator` parasitoid wasp (NCBI BioProject: PRJNA252202) using the hymenopteran lineage dataset the command would be:

```
$ python /path/to/busco/scripts/run_BUSCO.py
--in /path/to/wasp/transcriptomes/Pelecinus_polyturator.fs
--out PPOLY
--lineage /path/to/lineage/dataset/hymenoptera_odb9
--mode transcriptome
```

2. Running this assessment of the 35,969 `Pelecinus polyturator` transcripts using the hymenopteran lineage dataset on 4 CPUs with otherwise default options should take approximately 2 h and 40 min (*see* **Note 10**). For this analysis the transcriptome was not pre-processed to remove highly similar transcripts (*see* **Note 6**).

3.4 Understanding the Results

1. Successful assessments will each produce a simple summary results file that reports the full command used to launch the assessment (this is useful in order to be able to re-run the same analysis), as well as the percentages and counts of "*complete*" (single-copy and duplicated), "*fragmented*", and "*missing*" BUSCOs. For example, the results of the *Anopheles arabiensis* genome assembly assessment:

```
C:98.2%[S:98.1%,D:0.1%],F:0.7%,M:1.1%,n:2799
2750  Complete BUSCOs (C)
2746  Complete and single-copy BUSCOs (S)
4     Complete and duplicated BUSCOs (D)
19    Fragmented BUSCOs (F)
30    Missing BUSCOs (M)
2799  Total BUSCO groups searched
```

2. All three assessment modes will also produce a "`full_table`" file with classification results for each BUSCO, a "`missing_busco_list`" file with the missing BUSCOs, and a "`hmmer_output`" directory with the full results of the HMM searches. In addition, genome and transcriptome mode assessments will report the results of tBLASTn searches, and genome mode results include the details for all the Augustus gene predictions as well as Augustus training parameters and the nucleotide and protein sequences of the identified complete single-copy BUSCOs (*see* **Note 12**).
3. The BUSCO plotting tool enables users to visualize their results as a simple bar chart, allowing for clear comparisons of different datasets (*see* **Note 13**). To generate a chart, users must first copy the short summary results files from each assessment that they want to visualize into a single directory. The `generate_plot.py` script can then be launched pointing to this directory to automatically produce the chart. For example, the commands below will plot the results from the *Anopheles*

arabiensis genome (AARAD1) and gene set (AARAD16) assessments, producing the image file `busco_figure.png` in the same directory where the summary results were copied:

```
$ mkdir arabiensis_results
$ cp run_AARAD1/short_summary_AARAD1.txt arabiensis_results/.
$ cp run_AARAD16/short_summary_AARAD16.txt arabiensis_results/.
$ python /path/to/busco/scripts/generate_plot.py --wd arabiensis_results/
```

4. Repeating the steps outlined above to assess a total of 15 publicly available mosquito genome assemblies and their annotated gene sets and then plotting the results enables the like-for-like comparison of these genomic resources, where all but five datasets are more than 95% "*complete*" (Fig. 1). Furthermore, these mosquito genomics resources all show very low levels of duplications, indicating that the assemblies are likely mostly free of haplotype regions (*see* **Note 14**). In addition, the genome assembly results generally mirror those of the gene sets, with the assemblies usually performing slightly better apart from a few cases where the assembly appears substantially better than the gene set (*see* **Note 15**). These estimates of "*complete*", "*fragmented*", and "*missing*" BUSCOs (*see* **Note 16**) provide intuitive metrics with which to gauge the relative quality of these genomic resources in terms of their expected gene content.
5. Repeating the steps outlined above for the assessments of many more publicly available hymenopteran transcriptomes and plotting the resulting completeness scores against the numbers of transcripts demonstrates their highly variable completeness (Fig. 2). Transcriptomes may well be expected to show rather variable completeness scores as the total repertoire of RNAs that are sequenced and assembled will often reflect the type of biological sample, e.g., a pooled sample from multiple tissues and life-stages will probably capture more than a sample from a specialized tissue (*see* **Note 17**).

4 Notes

1. Users are encouraged to use the git client option to retrieve the BUSCO software as this will make installation of future updates much simpler and easy to manage. Additionally, the GitLab project "*issues*" page is worth consulting as it is a good source of tips and discussions from BUSCO users.
2. It has been reported that when running BUSCO using multiple cores, the tBLASTn step from BLAST+ versions 2.4, 2.5, and 2.6 may occasionally fail to complete and thus the BUSCO

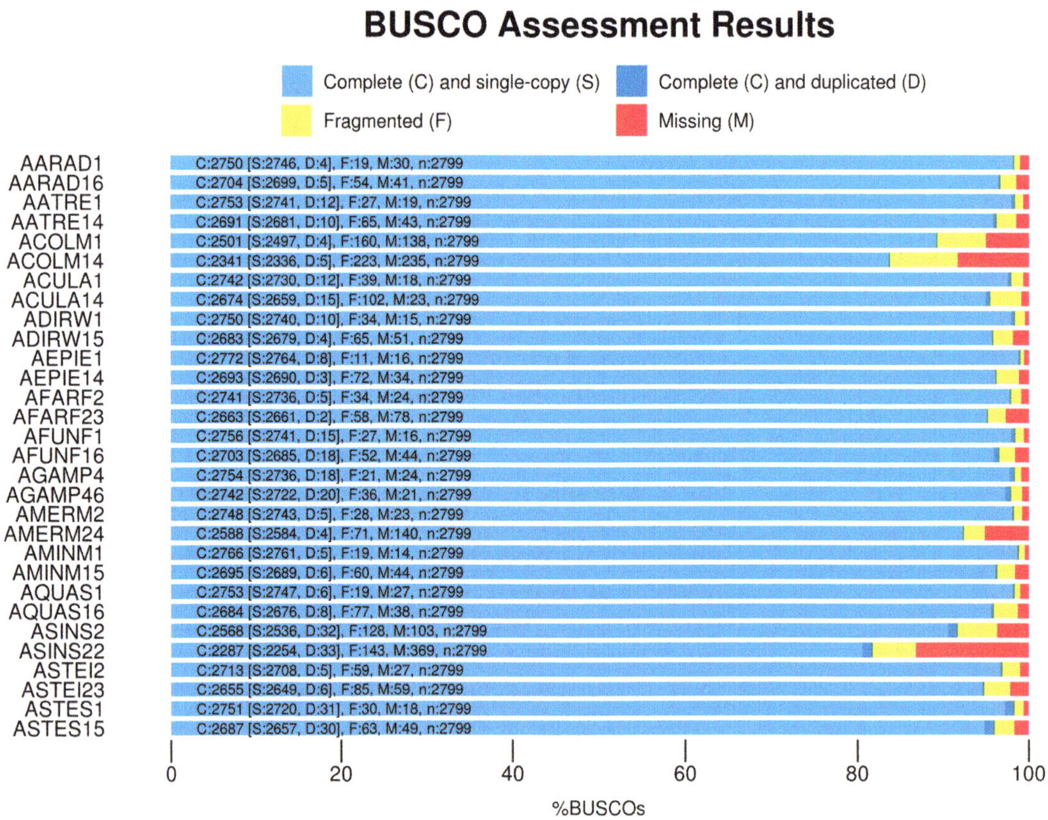

Fig. 1 BUSCO results from the assessments of 15 *Anopheles* mosquito genomes and their annotated gene sets. The chart was produced using the BUSCO plotting tool and demonstrates the intuitive visualization of side-by-side genome and gene set results from multiple species. Gene set results (e.g., AARAD16) are plotted directly beneath genome assembly results (e.g., AARAD1) for each species with numbers indicating dataset versions: *An. arabiensis* (AARAD), *An. atroparvus* (AATRE), *An. coluzzii* (ACOLM), *An. culicifacies* (ACULA), *An. dirus* (ADIRW), *An. epiroticus* (AEPIE), *An. farauti* (AFARF), *An. funestus* (AFUNF), *An. gambiae* (AGAMP), *An. merus* (AMERM), *An. minimus* (AMINM), *An. quadriannulatus* (AQUAS), *An. sinensis* (ASINS), *An. stephensi* Indian (ASTEI), *An. stephensi* SDA-500 (ASTES)

assessment will fail with an error message to this effect. To avoid this problem, use an earlier BLAST+ version or run using only a single core.

3. Users only need to install AUGUSTUS if they plan to assess genome assemblies. As AUGUSTUS has dependencies of its own, e.g., Perl, users should consult the AUGUSTUS documentation for the correct installation procedures. If working on a system where AUGUSTUS has already been installed by an administrator and the user does not have "write permission" to the AUGUSTUS "config" directory, users can simply recursively copy the entire "config" directory to a location where they do have "write permission" and then re-set the "config" path variable to this location:

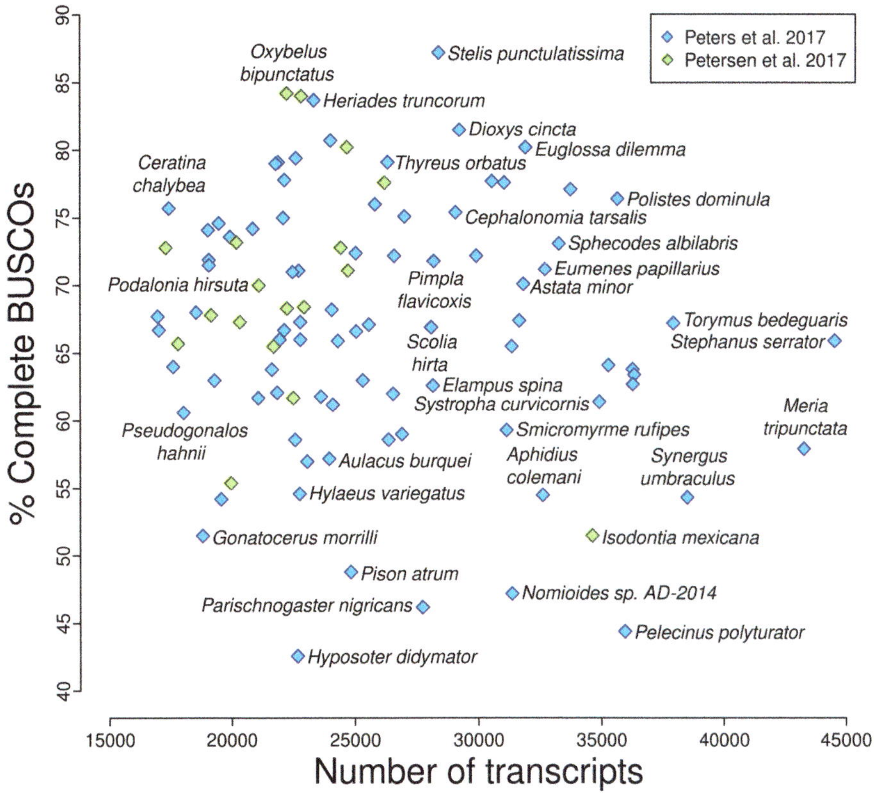

Fig. 2 BUSCO completeness results from the assessments of 103 hymenopteran transcriptomes from Peters et al. 2017 [24] and Petersen et al. 2017 [25] compared to the number of transcripts in each transcriptome. Transcriptomes with many transcripts are not necessarily the most complete, and those with fewer transcripts can still score relatively well in terms of completeness. Several example species are labeled either directly to the left or right of the data point or centered directly above or below it

```
$ cp -r /path/to/AUGUSTUS/augustus-3.2.3/config /my/home/augustus/config
$ export AUGUSTUS_CONFIG_PATH="/my/home/augustus/config/"
```

4. This was not a requirement for BUSCO v1 or v2. The v3 update refactored the underlying analysis code to make it more modular and extendable and thus it must be installed using the `setup.py` script.

5. For annotated gene sets the transcript-to-gene relationships are defined in the annotation files, e.g., General Feature Format (GFF) files, so the longest protein-coding transcript can be selected for each gene with multiple annotated transcripts. For *de novo* transcriptomes, i.e., those without a reference genome, transcript-to-gene relationships are not defined so users have two options: (a) run the assessments without pre-processing and acknowledge the fact that estimates of duplicated BUSCOs are likely to be inflated by the presence of multiple transcripts from the same gene, or (b) pre-process the

transcriptome with a sequence identity (or similarity) and length filter to select just one representative from sets of highly similar transcripts, e.g., using CD-HIT [26].

6. There are currently 16 bacterial lineage datasets and 28 eukaryotic lineage datasets. Users would normally select the most specific lineage available, i.e., the most recent ancestor of the species whose data is to be assessed. For example, for assessing ant data one would select the "`hymenoptera`" lineage rather than the "`arthropoda`" lineage. However, if there are a large number of species/strains/versions, etc. to be assessed, then to minimize runtime (at the expense of resolution) one might select a less specific (more ancestral) lineage dataset with fewer BUSCOs, at least for the initial rounds of assessments.
7. BUSCO outputs the running log details to the default standard output (user's terminal), in order to instead send these details to a file users can simply end the launch command with a redirect command, and as BUSCO assessments can take some time it is useful to run them in the background, i.e., end launch command with: `>& my_log_file.txt &`.
8. The optional arguments for launching a BUSCO assessment give the user flexibility over many aspects, some specific to running in genome mode and others applicable in any mode, all of which are described in full in the user guide. Some useful options to consider employing include (a) `--force`, this will force the results to overwrite results from an analysis run with the same name (b) `--tarzip`, this will package and compress the results from steps that can produce many output files; (c) `--augustus_options`, this allows users to pass AUGUSTUS-specific parameters for gene prediction, e.g., to use alternative codon translation tables.
9. AUGUSTUS comes with pretrained gene prediction parameters for many species (*see* AUGUSTUS documentation for up-to-date details), so if parameter sets are available for the species to be assessed, then they should be selected, e.g., for the Florida carpenter ant, the parameter set to use can be specified by adding the argument "`--species camponotus_floridanus`" to the launch command. For many other species, pretrained gene prediction parameters are not yet available, so users should select the closest species for which such parameters are available, or run the assessment with the preselected default parameters. For the sake of reproducibility it is important to specify which one was selected when reporting BUSCO results.
10. Assessment runtimes will vary according to the exact system setup. Assessments of genome assemblies require the initial steps of first identifying genomic regions that potentially

harbor BUSCO matches and then predicting gene models in these regions. These are computationally intensive tasks and therefore genome assembly assessments will take substantially longer than transcriptome or gene set assessments. Note also that the searches and gene predictions are performed in two rounds: (a) searches with consensus sequences built from BUSCO HMMs followed by gene predictions using the selected AUGUSTUS pretrained parameter set; (b) then for BUSCOs that were classified as "*fragmented*" or "*missing*" after the first round, searches with variant consensus sequences followed by gene predictions using parameters trained on "*complete*" BUSCOs identified in round one. Thus, if the first round identifies a high proportion of "*complete*" BUSCOs, then the second round will be relatively quick, but if there are many "*fragmented*" or "*missing*" BUSCOs after the first round, then the second round will take considerably longer.

11. The *Anopheles arabiensis* AaraD1.6 annotation contains 13,452 protein-coding genes with 13,640 transcripts so the protein FASTA file downloaded from VectorBase was first pre-processed to select the longest protein per gene. Performing this pre-processing step on annotated gene sets is not obligatory, but it ensures that BUSCO estimates of the numbers of duplicated genes will be true assessments that are not inflated by alternative transcripts that would be reported as gene duplicates.
12. During genome assessments the second round of gene predictions uses parameter sets built from the "*complete*" BUSCOs identified in the first round. These AUGUSTUS retraining parameters are saved in the "`augustus_output`" results directory. They are ideal for use during whole genome annotation procedures that employ AUGUSTUS, especially when parameter sets for the species to be annotated or those of a close relative are not already available. In addition, the GenBank or GFF formatted "*complete*" BUSCO annotations provided in the results directory can be used to train other gene predictors, e.g., SNAP [27].
13. The BUSCO plotting tool uses R (https://www.r-project.org) and the GGPLOT2 library (http://ggplot2.org), so these must be installed and accessible on the system in order to produce the image. Alternatively, adding the optional argument `--no_r` to the command will simply produce the R script required to build the image and users can then run this R script on any system where R and GGPLOT2 are installed. This also gives the user the opportunity to edit the R script to tailor the resulting image, e.g., changing the default fonts, labels, or colors.
14. If high levels of complete duplicates are reported for a genome assembly, then this could suggest that the assembly procedure

has failed to correctly collapse haplotype regions, resulting in numerous pairs of highly similar duplicate gene copies. This would warrant further investigations to determine if this is indeed the case and if alternative assembly strategies need to be employed or if such regions can be removed or collapsed. However, knowledge of the biology of the sample itself can also offer explanations: e.g., assessing the Aedes albopictus C6/36 cell line genome assembly and annotated gene set suggested that most BUSCOs were duplicated, but cytogenetic studies have shown that this cell line does have aberrant karyotypes, which could explain the numerous duplicates [28].

15. Differences in the results from assessing a genome assembly versus its annotated gene set may be due to several factors. In both cases BUSCO attempts to classify the matches to a set of protein-coding gene annotations: for genomes these annotations are built by BUSCO using AUGUSTUS gene predictions with BUSCO HMMs, whereas for gene sets they have usually been built by genome annotation pipelines (e.g., MAKER [29]) that incorporate evidence from several gene predictors and different sources of gene model support. Thus when results from assembly assessments appear to be better than for their gene sets it suggests that the targeted approach taken by BUSCO has produced better gene models than a more generalist annotation pipeline (at least for the subset of genes that make up the BUSCO lineage dataset). Conversely, if a gene set appears more complete than its genome, this suggests that the multiple sources of evidence used by the annotation pipeline have resulted in generally better annotations than the single-predictor approach taken by BUSCO.

16. When interpreting BUSCO results, users should be aware that the classification procedure (described in the introduction) results in the labels "*complete*", "*fragmented*", and "*missing*", which are by necessity simplifications that reflect the most likely scenario. For example, the label "*missing*" is applied to BUSCOs with no matches (probably truly absent from the dataset), but also to matches that do not meet the HMM score cutoffs. These below-cutoff matches could mean that these BUSCOs are in fact partially present in the dataset but there is simply not enough matching sequence be confident of the partial match and classify them as "*fragmented*".

17. BUSCO assessments are usually performed to demonstrate the good completeness levels of the genomic resources generated and analyzed in a particular study. However, if the aim of a transcriptomics experiment is to sample a specific tissue or life-stage where the repertoire of transcripts is expected to be highly specialized, then low completeness scores would in fact offer support that such targeted sampling was successful.

Acknowledgments

R.M.W. was supported by Swiss National Science Foundation award PP00P3_170664.

References

1. Richards S, Murali SC (2015) Best practices in insect genome sequencing: what works and what doesn't. Curr Opin Insect Sci 7:1–7
2. Gurevich A, Saveliev V, Vyahhi N et al (2013) QUAST: quality assessment tool for genome assemblies. Bioinformatics 29:1072–1075
3. Hunt M, Kikuchi T, Sanders M et al (2013) REAPR: a universal tool for genome assembly evaluation. Genome Biol 14:R47
4. Allen SL, Delaney EK, Kopp A et al (2017) Single-Molecule Sequencing of the *Drosophila serrata* Genome. G3: Genes, Genomes, Genetics 7:781–788
5. Davey JW, Chouteau M, Barker SL et al (2016) Major improvements to the *Heliconius melpomene* genome assembly used to confirm 10 chromosome fusion events in 6 million years of butterfly evolution. G3: Genes, Genomes, Genetics 6:695–708
6. Kanost MR, Arrese EL, Cao X et al (2016) Multifaceted biological insights from a draft genome sequence of the tobacco hornworm moth, *Manduca sexta*. Insect Biochem Mol Biol 76:118–147
7. Nowell RW, Elsworth B, Oostra V et al (2017) A high-coverage draft genome of the mycalesine butterfly *Bicyclus anynana*. GigaScience 6:1–7
8. Papanicolaou A, Schetelig MF, Arensburger P et al (2016) The whole genome sequence of the Mediterranean fruit fly, *Ceratitis capitata* (Wiedemann), reveals insights into the biology and adaptive evolution of a highly invasive pest species. Genome Biol 17:192
9. Benoit JB, Adelman ZN, Reinhardt K et al (2016) Unique features of a global human ectoparasite identified through sequencing of the bed bug genome. Nat Commun 7:10165
10. McKenna DD, Scully ED, Pauchet Y et al (2016) Genome of the Asian longhorned beetle (*Anoplophora glabripennis*), a globally significant invasive species, reveals key functional and evolutionary innovations at the beetle-plant interface. Genome Biol 17:227
11. Ioannidis P, Simão FA, Waterhouse RM et al (2017) Genomic features of the damselfly *Calopteryx splendens* representing a sister clade to most insect orders. Genome Biol Evol 9:415–430
12. Simão FA, Waterhouse RM, Ioannidis P et al (2015) BUSCO: Assessing genome assembly and annotation completeness with single-copy orthologs. Bioinformatics 31:3210–3212
13. Waterhouse RM, Seppey M, Simão FA et al (2017) BUSCO applications from quality assessments to gene prediction and phylogenomics. Mol Biol Evol 35:543–548
14. Zdobnov EM, Tegenfeldt F, Kuznetsov D et al (2017) OrthoDB v9.1: Cataloging evolutionary and functional annotations for animal, fungal, plant, archaeal, bacterial and viral orthologs. Nucleic Acids Res 45:D744–D749
15. Waterhouse RM, Zdobnov EM, Kriventseva EV (2011) Correlating traits of gene retention, sequence divergence, duplicability and essentiality in vertebrates, arthropods, and fungi. Genome Biol Evol 3:75–86
16. Waterhouse RM (2015) A maturing understanding of the composition of the insect gene repertoire. Curr Opin Insect Sci 7:15–23
17. Eddy SR (2011) Accelerated Profile HMM Searches. PLoS Comput Biol 7:e1002195
18. Keller O, Kollmar M, Stanke M et al (2011) A novel hybrid gene prediction method employing protein multiple sequence alignments. Bioinformatics 27:757–763
19. Camacho C, Coulouris G, Avagyan V et al (2009) BLAST+: architecture and applications. BMC Bioinformatics 10:421
20. Holt RA, Subramanian GM, Halpern A et al (2002) The genome sequence of the malaria mosquito *Anopheles gambiae*. Science 298:129–149
21. Jiang X, Peery A, Hall AB et al (2014) Genome analysis of a major urban malaria vector mosquito, *Anopheles stephensi*. Genome Biol 15:459
22. Neafsey DE, Waterhouse RM, Abai MR et al (2015) Highly evolvable malaria vectors: the genomes of 16 *Anopheles* mosquitoes. Science 347:1258522–1258522
23. Giraldo-Calderón GI, Emrich SJ, MacCallum RM et al (2015) VectorBase: an updated bioinformatics resource for invertebrate vectors and

other organisms related with human diseases. Nucleic Acids Res 43:D707–D713
24. Peters RS, Krogmann L, Mayer C et al (2017) Evolutionary history of the hymenoptera. Curr Biol 27:1013–1018
25. Petersen M, Meusemann K, Donath A et al (2017) Orthograph: a versatile tool for mapping coding nucleotide sequences to clusters of orthologous genes. BMC Bioinformatics 18:111
26. Li W, Godzik A (2006) CD-hit: a fast program for clustering and comparing large sets of protein or nucleotide sequences. Bioinformatics 22:1658–1659
27. Korf I (2004) Gene finding in novel genomes. BMC Bioinformatics 5:59
28. Waterhouse RM, Chen X, Bonizzoni M et al (2017) The third International Workshop on *Aedes albopictus*: building scientific alliances in the fight against the globally invasive Asian tiger mosquito. Pathog Global Health 111:161–165
29. Campbell MS, Holt C, Moore B et al (2014) Genome annotation and curation using MAKER and MAKER-P. Curr Protoc Bioinformatics 48:4.11.1-39

Chapter 7

The GFF3toolkit: QC and Merge Pipeline for Genome Annotation

Mei-Ju May Chen, Han Lin, Li-Mei Chiang, Christopher P. Childers, and Monica F. Poelchau

Abstract

The GFF3toolkit (https://github.com/NAL-i5K/GFF3toolkit) supported by the i5k Workspace@NAL provides a suite of tools to handle gene annotations in GFF3 format from arthropod genome projects and their research communities. To improve GFF3 formatting of gene annotations, a quality control and merge procedure is proposed along with the GFF3toolkit. In particular, the toolkit provides functions to sort a GFF3 file, detect GFF3 format errors, merge two GFF3 files, and generate biological sequences from a GFF3 file. This chapter explains when and how to use the provided tools to obtain nonredundant arthropod gene sets in high quality.

Key words I5k, Arthropods, Insects, Genomics, Community annotation, Gene annotations, GFF3

1 Introduction

Advances in sequencing technology have opened a new era in genomics, and have changed approaches to gene annotation. Some studies have revealed that community genome annotation distills invaluable knowledge for genome projects [1, 2]. Recently, several insect genomes sequenced under the purview of the i5k consortium were collectively annotated by the scientific community, under the stewardship of the National Agricultural Library's i5k Workspace@NAL (https://i5k.nal.usda.gov) [3–5]. However, harnessing the results of community annotation efforts can be fraught with difficulties, including adherence to basic formatting standards.

The i5k Workspace@NAL is a genome portal established to facilitate access, visualization, and curation for insect and arthropod genomes [6]. One of its roles is to guide community annotators through the manual annotation process in order to generate improved reference gene sets for the broader scientific community. Any public arthropod

Susan J. Brown and Michael E. Pfrender (eds.), *Insect Genomics: Methods and Protocols*, Methods in Molecular Biology, vol. 1858, https://doi.org/10.1007/978-1-4939-8775-7_7,

genome that is accessioned in the INSDC (International Nucleotide Sequence Database Collaboration) can submit their genome project to the i5k Workspace and take advantage of its resources and support for manual annotation. To help these scientific communities involved in insect genome projects easily process gene annotations in GFF3 (General Feature Format Version 3; https://github.com/The-Sequence-Ontology/Specifications/blob/master/gff3.md), we have developed the GFF3toolkit (https://github.com/NAL-i5K/GFF3toolkit), a quality control (QC) and merge pipeline for GFF3 files.

GFF3 is one of the standard formats used to describe and represent genomic features. It is an incredibly flexible, 9-column text format, which is easily manipulated by biologists. This flexibility, however, makes it very easy to violate the GFF3 specification (http://www.sequenceontology.org/), which can result in problems when using downstream programs. General formatting errors can be found by GFF3 validation software, such as GAL (https://github.com/The-Sequence-Ontology/GAL) or GFF3 validator (http://genometools.org/cgi-bin/gff3validator.cgi) [7]. These tools aim to identify gene models that violate the GFF3 format specification. The GFF3toolkit package additionally recognizes GFF3 formatting errors that may not violate the GFF3 specification, but may nonetheless be mis-representations of a gene model. For example, in Fig. 1, two annotators (A and B) annotate two genes in the same genomic region, which share coding sequence with each other. Biologically speaking, these two genes are likely two isoforms of the same gene. However, in the resulting GFF3 file, these models are represented as two separate gene models (Gene 1 and Gene 2), rather than two isoforms of the same gene. To tackle this issue, the GFF3toolkit provides a QC procedure that aims to detect over 50 types of formatting errors, including those arising from community annotation.

Another important task during annotation projects is the merging of two GFF3 files into a nonredundant gene set. For example, the i5k Workspace@NAL hosts more than 55 insect genome projects and uses the manual annotation program Apollo (http://genomearchitect.github.io/) [8] to perform community annotation. For each project, community annotation can generate thousands of revised or newly added models that need to be incorporated into a reference gene set. To meet this need, a merge procedure was developed to integrate output from Apollo with a single reference GFF3 file to generate an updated gene set. For a manually curated model, the merge procedure can

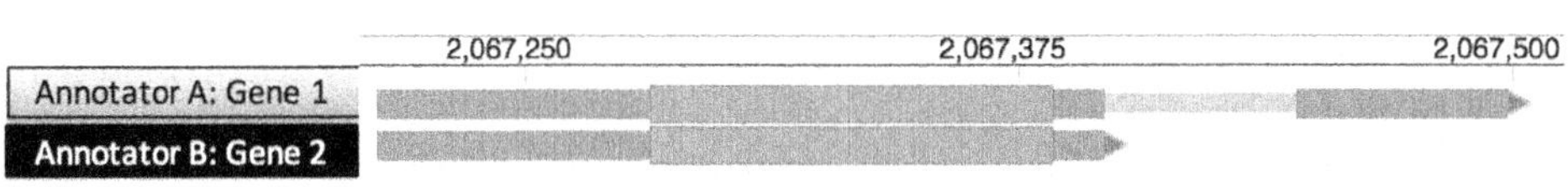

Fig. 1 A typical example of formatting errors arising from community annotation—incorrectly separate genes

automatically identify models in the reference gene set that should be replaced, and then perform the replacement. The procedure involves a custom GFF3 attribute in the manually curated GFF3 file—the replace tag—that specifies which mRNA(s) or transcript (s) from a single reference GFF3 file should be replaced by the new annotation. This method allows some flexibility in the merge procedure: users can add replace tags manually if needed, or can use other tools, such as ParsEval [9] or BEDTools [10], to identify which reference models should be replaced.

This chapter provides guidance for using version 1.1 of the program suite to process and generate high-quality genome annotations. Any questions about the GFF3toolkit can be directed to i5k@ars.usda.gov, or the issue page of the NAL-i5K GitHub repository: https://github.com/NAL-i5K/GFF3toolkit/issues.

2 Materials

2.1 Package Download

The GFF3toolkit version1.1.0 can be downloaded from this link: https://github.com/NAL-i5K/GFF3toolkit/releases/tag/v1.1.0. Those wishing to take advantage of GitHub's features can clone the repository from https://github.com/NAL-i5K/GFF3toolkit/tree/v1.1.0, using the download link as shown in Fig. 2 (this requires a GitHub account). Users can download the program suite as a compressed file (.zip), or by the git command: "git clone https://github.com/NAL-i5K/GFF3toolkit.git," followed by "git checkout tags/v1.1.0." For more information on how to use GitHub, refer to GitHub Guides (https://guides.github.com/).

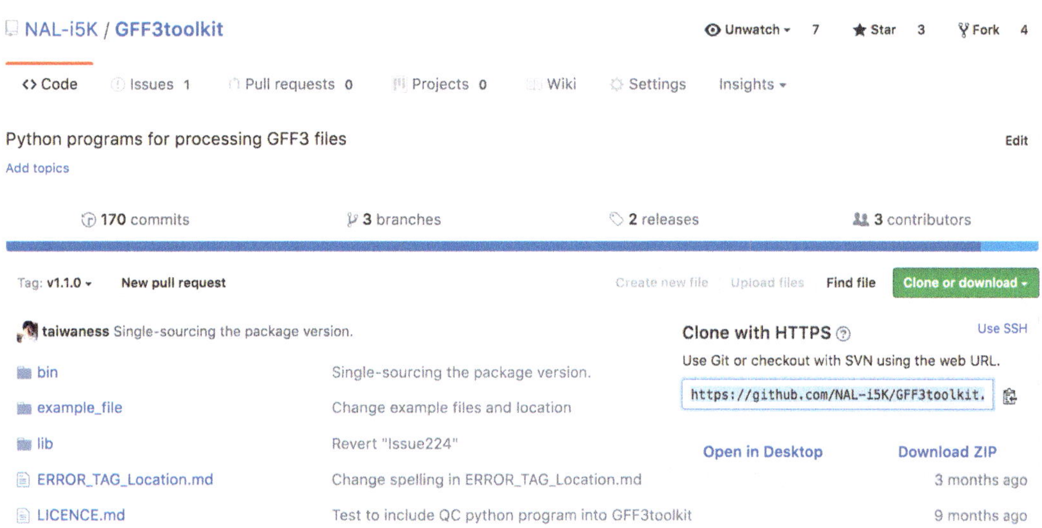

Fig. 2 The GitHub repository of the GFF3toolkit. Click the green "Clone or download" box to show multiple ways to acquire the program suite

2.2 Package Dependencies that Need to be Installed in Advance

1. Python 2.7 (https://www.python.org/downloads/).
2. Perl v5.16.3 (https://www.perl.org/get.html).

2.3 Input Files

1. New or modified gene annotations in GFF3 format (GFF3 file 1), to be merged into the reference gene set (*see* **Notes 1** and **2**).
2. Reference gene models in GFF3 format (GFF3 file 2). The models from GFF3 file 1 will be merged into this file, replacing models in GFF3 file 2 (*see* **Notes 1** and **2**).
3. Genome sequences in FASTA format. This file must be the FASTA file that the GFF3 sequence IDs and coordinates in both GFF3 files refer to (*see* **Note 2**).

3 Methods

3.1 Functions and the Corresponding Programs Provided by the GFF3toolkit

We have developed the GFF3toolkit to:

1. Sort GFF3 files, which can aid downstream processing and custom parsing (*gff3_sort.py*).
2. Help identify common problems with GFF3 files (*gff3_QC.py*).
3. Merge two GFF3 files into a single, nonredundant GFF3 file (*gff3_merge.py*).
4. Generate FASTA files from a GFF3 file for many use cases, e.g., feature types beyond mRNA (*gff3_to_fasta.py*).

The programs for the above four functions can be found in the **GFF3toolkit/bin** directory (Fig. 2). All of the programs can be independently executed for custom use. Nevertheless, we highly recommend the QC and Merge pipeline (Fig. 3) utilizing the GFF3toolkit to acquire high-quality gene annotations. Example files (in the **GFF3toolkit/example_file** directory) and example commands (https://github.com/NAL-i5K/GFF3toolkit/blob/master/README.md) are provided to familiarize users with the input formats and the program suite.

3.2 Quality Control and Merge Pipeline

Here, we describe the QC and Merge pipeline (Fig. 3) step by step.

1. **Sort input GFF3 files.** Before running the pipeline, we suggest sorting features in the input GFF3 files by Scaffold, coordinates on a Scaffold, and parent-child feature relationships. Both manually curated (GFF3 file 1) and reference (GFF3 file 2) models should be sorted, respectively (*see* **gff3_sort** in Fig. 3). The GFF3toolkit provides *gff3_sort.py* for sorting and reformatting GFF3 files (*see* **Note 3**).

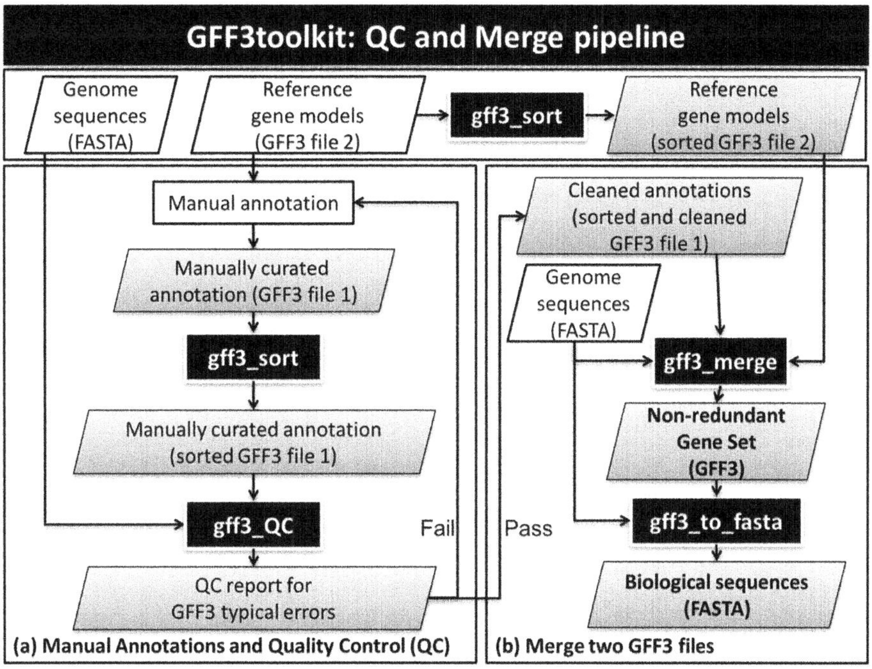

Fig. 3 The GFF3toolkit QC and Merge pipeline. Black rectangles represent the programs from the GFF3toolkit. Gray parallelograms are the output files generated by those programs. White parallelograms and rectangles indicate input files or the program from another source. (**a**) steps required or suggested for manual annotations and quality control; (**b**) steps required or suggested for merging two GFF3 files

Program: ***gff3_sort.py***.
Inputs: a GFF3 file (GFF3 file 1 or GFF3 file 2).
Output: a sorted GFF3 file.
ReadMe: https://github.com/NAL-i5K/GFF3toolkit/blob/master/docs/gff3_sort.md

2. **Detect GFF3 format errors (QC phase; Fig. 3(a)).** The ***gff3_QC.py*** program detects over 50 types of formatting errors listed in Table 1 (*see* **Note 4**). Error types are grouped into three categories:
   (a) Intra-model errors (Ema)—Formatting errors that can be found by jointly considering multiple features within a gene model, such as gene, mRNA, exon, and CDS features. Errors in this category are given an "Error_Code" starting with "Ema."
   (b) Inter-model errors (Emr)—Formatting errors that can be found by comparing multiple gene models. Errors in this category are given an "Error_Code" starting with "Emr."
   (c) Single feature errors (Esf)—Formatting errors that can be found by searching the GFF3 file line by line. Errors in this category are given an "Error_Code" starting with "Esf."

Table 1
All errors currently considered by *gff3_QC.py*, including the error code and the error tag (a brief explanation of the error)

Error_Code	Error_Tag
Ema0001	Parent feature start and end coordinates exceed those of child features
Ema0002	Protein sequence contains internal stop codons
Ema0003	This feature is not contained within the parent feature coordinates
Ema0004	Incomplete gene feature that should contain at least one mRNA, exon, and CDS
Ema0005	Pseudogene has invalid child feature type
Ema0006	Wrong phase
Ema0007	CDS and parent feature on different strands
Ema0008	Warning for distinct isoforms that do not share any regions
Ema0009	Incorrectly merged gene parent? Isoforms that do not share coding sequences are found
Emr0001	Duplicate transcript found
Emr0002	Incorrectly split gene parent?
Emr0003	Duplicate ID
Esf0001	Feature type may need to be changed to pseudogene
Esf0002	Start/Stop is not a valid 1-based integer coordinate
Esf0003	strand information missing
Esf0004	Seqid not found in any ##sequence-region
Esf0005	Start is less than the ##sequence-region start
Esf0006	End is greater than the ##sequence-region end
Esf0007	Seqid not found in the embedded ##FASTA
Esf0008	End is greater than the embedded ##FASTA sequence length
Esf0009	Found Ns in a feature using the embedded ##FASTA
Esf0010	Seqid not found in the external FASTA file
Esf0011	End is greater than the external FASTA sequence length
Esf0012	Found Ns in a feature using the external FASTA
Esf0013	White chars not allowed at the start of a line
Esf0014	##gff-version missing from the first line
Esf0015	Expecting certain fields in the feature
Esf0016	##sequence-region seqid may only appear once
Esf0017	Start/End is not a valid integer
Esf0018	Start is not less than or equal to end
Esf0019	Version is not "3"

(continued)

Table 1
(continued)

Error_Code	Error_Tag
Esf0020	Version is not a valid integer
Esf0021	Unknown directive
Esf0022	Features should contain 9 fields
Esf0023	escape certain characters
Esf0024	Score is not a valid floating point number
Esf0025	Strand has illegal characters
Esf0026	Phase is not 0, 1, or 2, or not a valid integer
Esf0027	Phase is required for all CDS features
Esf0028	Attributes must escape the percent (%) sign and any control characters
Esf0029	Attributes must contain one and only one equal (=) sign
Esf0030	Empty attribute tag
Esf0031	Empty attribute value
Esf0032	Found multiple attribute tags
Esf0033	Found "," in a attribute, possible unescaped
Esf0034	attribute has identical values (count, value)
Esf0035	attribute has unresolved forward reference
Esf0036	Value of an attribute contains unescaped ","
Esf0037	Target attribute should have 3 or 4 values
Esf0038	Start/End value of Target attribute is not a valid integer coordinate
Esf0039	Strand value of Target attribute has illegal characters
Esf0040	Value of Is_circular attribute is not "true"
Esf0041	Unknown reserved (uppercase) attribute

The ***gff3_QC.py*** program can be applied on both manually curated (GFF3 file 1) and reference (GFF3 file 2) files (*see* **Note 5**). At a minimum, we recommend running the program on the manually curated GFF3 file.

Program: ***gff3_QC.py***.

Inputs: a GFF3 file (at least GFF3 file 1), and the FASTA of genome sequences that the GFF3 sequence IDs and coordinates refer to.

Output: a tab-delimited text file of QC report.

ReadMe: https://github.com/NAL-i5K/GFF3toolkit/blob/master/docs/gff3_QC.md

3. **Check QC reports and correct problematic models (QC phase; Fig. 3(a)).** If there are errors found in the QC report of a GFF3 file, the GFF3 file needs to be reviewed and corrected (*see* the **Manual annotation** box in Fig. 3a). Currently, problematic gene models identified via *gff3_QC.py* have to be manually corrected. We recommend using the manual annotation program Apollo to assist in the review and correction process when possible. However, if users work on a genome alone or with a small group of people, the users can directly edit the curated GFF3 file (GFF3 file 1 or the sorted GFF3 file 1) instead of using Apollo. Repeat **steps 2** and **3** until GFF3 files (at least the manually curated GFF3 file) are as clean as possible. Otherwise, the subsequent Merge phase may fail or incorporate errors into the merged GFF3 files.
4. **Merge two GFF3 files (Merge phase; Fig. 3b).** The program *gff3_merge.py* can be conceptually separated into three steps:
   (a) Recognize or auto-assign Replace Tags (*see* **Note 6**) to transcripts or mRNAs in the manually curated GFF3 file.
   (b) Determine merge actions based on the Replace Tags:
   Deletion—a model has the status attribute "Delete."
   Simple replacement—a model has a single replace tag.
   New addition—a model has a replace tag "NA."
   Split replacement—a modified model shares a replace tag with other modified models.
   Merge replacement—a model has multiple replace tags.
   (c) Models from a manually curated GFF3 file replace models from a reference GFF3 file based on merge actions in **step 4b**. *See* **Note** 7.

In this Merge phase, you can choose to have the program auto-assign replace tags for you (https://github.com/NAL-i5K/GFF3toolkit/wiki/Merge-two-GFF3-files#automatically-assigning-replace-tags; *see* **Note 8**). This is the default behavior. Alternatively, you can also choose to add replace tags into the manually curated GFF3 file (the sorted and cleaned GFF3 file 1) on your own. However, unusual use cases may happen sometimes; please review **Note 9** prior to running the program. For more information about how the merge works, refer to the GFF3toolkit wiki: https://github.com/NAL-i5K/GFF3toolkit/wiki/Merge-two-GFF3-files.

Program: *gff3_merge.py*.

Inputs: two GFF3 files (sorted and cleaned GFF3 file 1 and GFF3 file 2), and the FASTA of genome sequences that the GFF3 sequence IDs and coordinates refer to.

Output: the merged GFF3 file of a nonredundant gene set.
ReadMe: https://github.com/NAL-i5K/GFF3toolkit/blob/master/docs/gff3_merge.md

5. **Generate biological sequences from a GFF3 file (Merge phase; Fig. 3(b))**. The ***gff3_to_fasta.py*** program provides a method to extract biological sequences from a GFF3 file. You can choose to extract one feature type from the six types that the program considers, or extract all six types at once. For advanced users, there is the additional option to specify your own sequence type that you'd like to extract by giving the program the corresponding parent and child features:
   (a) "all"—FASTA files for all types of sequences listed below.
   (b) "gene"—gene sequence for each record.
   (c) "exon"—exon sequence for each record.
   (d) "pre_trans"—genomic region of a transcript model (premature transcript).
   (e) "trans"—spliced transcripts (only exons included).
   (f) "cds"—coding sequences.
   (g) "pep"—peptide sequences.
   (h) "user_defined"—specify parent and child features via the -u argument.

For peptide sequences, translation (*see* **Note 10**) from 64 combinations of standard codons and universal stop codons is considered. However, sequence extraction based on a GFF3 file might generate incorrect sequences, if the GFF3 file contains problematic models. This program can automatically validate GFF3 format and list the detected errors for users. This is the default behavior. The QC function, which is an abbreviated version of gff3_QC.py, can be turned off by adding the option -noQC when running the program.
Program: ***gff3_to_fasta.py***.
Inputs: a GFF3 file (such as the merged GFF3 file), and the FASTA of genome sequences that the GFF3 sequence IDs and coordinates refer to.
Output: biological sequences in FASTA format.
ReadMe: https://github.com/NAL-i5K/GFF3toolkit/blob/master/docs/gff3_to_fasta.md

4 Notes

1. In this example workflow (Fig. 3), GFF3 file 1 serves as the input for ***gff3_sort.py, gff3_QC.py and gff3_merge.py***, while GFF3 file 2 is used with ***gff3_sort.py*** and ***gff3_merge.py***. Please

note that the ***gff3_merge.py*** and ***gff3_QC.py*** programs require gene, pseudogene, mRNA, and pseudogenic transcript features to have an ID attribute in column 9 of the GFF3 files.

2. For more information on the GFF3 format, refer to the GFF3 specification (https://github.com/The-Sequence-Ontology/Specifications/blob/master/gff3.md).
3. The program ***gff3_sort.py*** ignores hash tag lines other than **##gff-version 3** and **###**. After sorting, the program separates gene models using the **###** directive in the output GFF3. This may change the number of lines between the input and output file. Additionally, if a feature has two or more parent IDs, the program replicates the feature and lists it under each parent, resulting in an output file with more lines than the input file.
4. The ***gff3_QC.py*** program works best for protein-coding genes in "canonical" Sequence ontology style (*see* GFF3 specification: https://github.com/The-Sequence-Ontology/Specifications/blob/master/gff3.md). New functionality for noncoding features will be added in the future.
5. The ***gff3_QC.py*** program can take a while if the input GFF3 file is large, especially for reference GFF3 files.
6. The replace tag is a custom GFF3 attribute in GFF3 file 1 that specifies which mRNA(s) or transcript(s) from a single reference GFF3 file (GFF3 file 2) should be replaced by the new annotation. The replace tag format is: replace=[Name or ID attribute of reference mRNA or transcript to be replaced]. For examples of models with replace tags, refer to the GFF3toolkit wiki: https://github.com/NAL-i5K/GFF3toolkit/wiki/Merge-two-GFF3-files#replace-tags. Additionally, the replace tag could be added through the Apollo program (check the tutorial from the i5k Workspace@NAL: https://i5k.nal.usda.gov/apollo-replaced-models-field-explanations-and-examples) or directly added into the manually curated GFF3 file.
7. During the Merge phase, the manually curated model fully replaces the corresponding reference model in the merged GFF3 file. This means that even functional information from the replaced reference model, such as Name, symbols, etc., will not be retained.
8. The auto-assignment of replace tags ONLY works for mRNA features. For all other feature types, if there is no replace tag, the program will add "replace=NA," meaning that no models from the reference gff3 file will be replaced.
9. Unusual use cases:
   (a) Noncoding features: The merge program currently doesn't auto-assign replace tags for noncoding features—if no replace tag is present for a noncoding feature,

then the program will add "replace=NA," which means that no reference model will be replaced. If you would like a noncoding model to replace a model in the reference GFF3 file, then you must add the replace tag manually.

(b) Isoforms that don't share coding sequence: It is often assumed that isoforms of a gene model should contain some shared coding sequence. However, it is possible for gene models to have multiple isoforms that do not share CDS with each other—for example, partial gene models due to a poor genome assembly. In this case, the auto-assignment program will exhibit confusing behavior: it will assign different replace tags to each isoform, but will then reject these auto-assigned replace tags because it expects isoforms of a gene model to have the same replace tags. You will need to add the replace tags manually, such that all isoforms of a gene model carry the same replace tags. For example, if isoform A1 replaces model B, and isoform A2 replaces model C, then both isoform A1 and A2 should have replace tags "replace=B,C."

(c) New and reference model overlap in noncoding regions: If a new model overlaps a reference model in an intron or UTR (but not within the coding sequence), the auto-assignment program will NOT assign a replace tag. This is because it is not always clear whether the overlapping model should be replaced, and this case usually requires manual review. The program's standard error will display a warning message that this model does not have a replace tag and therefore was not incorporated into the merged GFF3 file. You can then go back and manually add a replace tag to the original GFF3 file.

10. A **translator** python method is provided by the ***gff3_to_fasta.py*** program. The method supports translation from 64 combinations of standard codons and universal stop codons, codons with IUB Depiction (https://www-bimas.cit.nih.gov/molbio/translate/codes.html), and mRNA (with U) or CDS (with T instead of U).

5 EEO/Nondiscrimination Statement

The U.S. Department of Agriculture (USDA) prohibits discrimination in all its programs and activities on the basis of race, color, national origin, age, disability, and where applicable, sex, marital status, familial status, parental status, religion, sexual orientation, genetic information, political beliefs, reprisal, or because all or part of an individual's income is derived from any public assistance program. (Not all prohibited bases apply to all programs.) Persons with disabilities who require alternative means for communication of program information (Braille, large print, audiotape, etc.) should contact USDA's TARGET Center at (202) 720-2600 (voice and TDD). To file a complaint of discrimination, write to USDA, Director, Office of Civil Rights, 1400 Independence Avenue, S. W., Washington, D.C. 20250-9410, or call (800) 795-3272 (voice) or (202) 720-6382 (TDD). USDA is an equal opportunity provider and employer.

Acknowledgments

We would like to thank Chien-Yueh Lee and Yu-Yu Lin for their suggestions on the early development of the program suite. Dan Hughes and Stephen (fringy) Richards had the initial idea for the "replace" tag for the merge program. Funding for this project is from the United States Department of Agriculture–Agricultural Research Service; and the USDA-ARS Bee Research Laboratory.

References

1. Wang Q, Arighi CN, King BL, Polson SW, Vincent J, Chen C, Huang H, Kingham BF, Page ST, Rendino MF, Thomas WK, Udwary DW, Wu CH, North East Bioinformatics Collaborative Curation T (2012) Community annotation and bioinformatics workforce development in concert—Little Skate Genome Annotation Workshops and Jamborees. Database (Oxford) 2012:bar064. https://doi.org/10.1093/database/bar064
2. Mazumder R, Natale DA, Julio JA, Yeh LS, Wu CH (2010) Community annotation in biology. Biol Direct 5:12. https://doi.org/10.1186/1745-6150-5-12
3. Benoit JB, Adelman ZN, Reinhardt K, Dolan A, Poelchau M, Jennings EC, Szuter EM, Hagan RW, Gujar H, Shukla JN, Zhu F, Mohan M, Nelson DR, Rosendale AJ, Derst C, Resnik V, Wernig S, Menegazzi P, Wegener C, Peschel N, Hendershot JM, Blenau W, Predel R, Johnston PR, Ioannidis P, Waterhouse RM, Nauen R, Schorn C, Ott MC, Maiwald F, Johnston JS, Gondhalekar AD, Scharf ME, Peterson BF, Raje KR, Hottel BA, Armisen D, Crumiere AJ, Refki PN, Santos ME, Sghaier E, Viala S, Khila A, Ahn SJ, Childers C, Lee CY, Lin H, Hughes DS, Duncan EJ, Murali SC, Qu J, Dugan S, Lee SL, Chao H, Dinh H, Han Y, Doddapaneni H, Worley KC, Muzny DM, Wheeler D, Panfilio KA, Vargas Jentzsch IM, Vargo EL, Booth W, Friedrich M, Weirauch MT, Anderson MA, Jones JW, Mittapalli O, Zhao C, Zhou JJ, Evans JD, Attardo GM, Robertson HM, Zdobnov EM, Ribeiro JM, Gibbs RA, Werren JH, Palli SR, Schal C, Richards S (2016) Unique features of a global human ectoparasite identified through sequencing of the bed bug genome. Nat Commun 7:10165. https://doi.org/10.1038/ncomms10165
4. McKenna DD, Scully ED, Pauchet Y, Hoover K, Kirsch R, Geib SM, Mitchell RF, Waterhouse RM, Ahn SJ, Arsala D, Benoit JB, Blackmon H, Bledsoe T, Bowsher JH, Busch A, Calla B, Chao H, Childers AK, Childers C, Clarke DJ, Cohen L, Demuth JP, Dinh H, Doddapaneni H, Dolan A, Duan JJ, Dugan S, Friedrich M, Glastad KM, Goodisman MA, Haddad S, Han Y, Hughes DS, Ioannidis P, Johnston JS, Jones JW, Kuhn LA, Lance DR, Lee CY, Lee SL, Lin H, Lynch JA, Moczek AP, Murali SC, Muzny DM, Nelson DR, Palli SR, Panfilio KA, Pers D, Poelchau MF, Quan H, Qu J, Ray AM, Rinehart JP, Robertson HM, Roehrdanz R, Rosendale AJ, Shin S, Silva C, Torson AS, Jentzsch IM, Werren JH, Worley KC, Yocum G, Zdobnov EM, Gibbs RA, Richards S (2016) Genome of the Asian longhorned beetle (*Anoplophora glabripennis*), a globally significant invasive species, reveals key functional and evolutionary innovations at the beetle-plant interface. Genome Biol 17(1):227. https://doi.org/10.1186/s13059-016-1088-8
5. Papanicolaou A, Schetelig MF, Arensburger P, Atkinson PW, Benoit JB, Bourtzis K,

Castanera P, Cavanaugh JP, Chao H, Childers C, Curril I, Dinh H, Doddapaneni H, Dolan A, Dugan S, Friedrich M, Gasperi G, Geib S, Georgakilas G, Gibbs RA, Giers SD, Gomulski LM, Gonzalez-Guzman M, Guillem-Amat A, Han Y, Hatzigeorgiou AG, Hernandez-Crespo P, Hughes DS, Jones JW, Karagkouni D, Koskinioti P, Lee SL, Malacrida AR, Manni M, Mathiopoulos K, Meccariello A, Murali SC, Murphy TD, Muzny DM, Oberhofer G, Ortego F, Paraskevopoulou MD, Poelchau M, Qu J, Reczko M, Robertson HM, Rosendale AJ, Rosselot AE, Saccone G, Salvemini M, Savini G, Schreiner P, Scolari F, Siciliano P, Sim SB, Tsiamis G, Urena E, Vlachos IS, Werren JH, Wimmer EA, Worley KC, Zacharopoulou A, Richards S, Handler AM (2016) The whole genome sequence of the Mediterranean fruit fly, *Ceratitis capitata* (Wiedemann), reveals insights into the biology and adaptive evolution of a highly invasive pest species. Genome Biol 17(1):192. https://doi.org/10.1186/s13059-016-1049-2

6. Poelchau M, Childers C, Moore G, Tsavatapalli V, Evans J, Lee CY, Lin H, Lin JW, Hackett K (2015) The i5k Workspace@-NAL--enabling genomic data access, visualization and curation of arthropod genomes. Nucleic Acids Res 43(Database issue): D714–D719. https://doi.org/10.1093/nar/gku983
7. Gremme G, Steinbiss S, Kurtz S (2013) GenomeTools: a comprehensive software library for efficient processing of structured genome annotations. IEEE/ACM Trans Comput Biol Bioinform 10(3):645–656. https://doi.org/10.1109/TCBB.2013.68
8. Lee E, Helt GA, Reese JT, Munoz-Torres MC, Childers CP, Buels RM, Stein L, Holmes IH, Elsik CG, Lewis SE (2013) Web Apollo: a web-based genomic annotation editing platform. Genome Biol 14(8):R93. https://doi.org/10.1186/gb-2013-14-8-r93
9. Standage DS, Brendel VP (2012) ParsEval: parallel comparison and analysis of gene structure annotations. BMC Bioinformatics 13:187. https://doi.org/10.1186/1471-2105-13-187
10. Quinlan AR, Hall IM (2010) BEDTools: a flexible suite of utilities for comparing genomic features. Bioinformatics 26(6):841–842. https://doi.org/10.1093/bioinformatics/btq033

Chapter 8

Using Formaldehyde-Assisted Isolation of Regulatory Elements (FAIRE) to Identify Functional Regulatory DNA in Insect Genomes

Daniel J. McKay

Abstract

Differential regulation of gene expression determines cell-type-specific function, making identification of the *cis*-regulatory elements that control gene expression a central goal of developmental biology. In addition, changes in the sequence of *cis*-regulatory elements are thought to drive changes in gene expression patterns between species, making comparisons of *cis*-regulatory element usage important for evolutionary biology as well. Due to the number of extant species and the incredible morphological diversity that they exhibit, insects are favorite model organisms for both developmental and evolutionary biologists alike. However, identifying *cis*-regulatory elements in insect genomes is challenging. Here, I describe a method termed FAIRE-seq (Formaldehyde-Assisted Isolation of Regulatory Elements, followed by high-throughput sequencing) that can be used to identify functional DNA regulatory elements from developing insect tissues, genome-wide.

Key words Open chromatin, *Cis*-regulatory element, Enhancer, FAIRE, *Drosophila* development

1 Introduction

Identifying *cis*-regulatory elements in insect genomes faces multiple challenges. As in other organisms, *cis*-regulatory elements such as enhancers and promoters are small, typically around 1000 base pairs in length. Enhancers in particular can also function over long distances and in a position-independent manner, meaning that they lack a stereotypical location relative to the genes they regulate. And in contrast to protein-coding genes, *cis*-regulatory elements lack a universal genetic code. For these reasons, accurate prediction of *cis*-regulatory elements based on DNA sequence alone is difficult.

A variety of experimental methods have been employed to directly identify *cis*-regulatory elements in target cells. Due to their small size, collecting sufficient amounts of input cells for genomic assays can be challenging in insects, especially if cell-

Susan J. Brown and Michael E. Pfrender (eds.), *Insect Genomics: Methods and Protocols*, Methods in Molecular Biology, vol. 1858, https://doi.org/10.1007/978-1-4939-8775-7_8, © Springer Science+Business Media, LLC, part of Springer Nature 2019

type-specific information is desired. Likewise, many insects undergo rapid changes during development, limiting the time available to identify *cis*-regulatory elements with temporal-specific activity. Here, I present a method termed FAIRE-seq (Formaldehyde-Assisted Isolation of Regulatory Elements) [1] that has successfully been used to identify functional *cis*-regulatory elements in insects, genome-wide. FAIRE-seq is a simple method that is sensitive enough to be used with limiting amounts of input cells. It also does not rely on enzymes or antibodies, thus simplifying the ability to identify and compare *cis*-regulatory element usage between species. Finally, since FAIRE-seq relies on formaldehyde-fixed samples, it allows time for precise tissue dissections and collection of short-lived developmental stages.

FAIRE-seq identifies sites of "open" chromatin in the genome [1–6]. Due to the competition between transcription factors and nucleosomes for DNA binding, active *cis*-regulatory elements, which are typically bound by transcription factors, are often depleted of nucleosomes or contain nucleosomes that are remodeled in such a way as to decrease their interactions with DNA [7]. In the FAIRE-seq protocol, samples are briefly fixed with formaldehyde, sonicated to shear chromatin into smaller fragments, and then subjected to a phenol-chloroform extraction to isolate water-soluble DNA, which is then prepared for high-throughput sequencing. FAIRE is thought to work by exploiting the differences between histones and transcription factors for being crosslinked to DNA. Due to the large number of amino acid residues that react with formaldehyde, and because of the extensive contacts that they make with DNA, histones are more likely to be crosslinked to DNA than transcription factors. As a result, genomic sites bound by transcription factors will preferentially segregate to the aqueous phase during the phenol-chloroform extraction, whereas DNA crosslinked to protein will be trapped at the interface between the aqueous and organic phases.

The ease of performing FAIRE-seq has led to its use with a range of developmental samples, including whole embryos [8], dissected tissues [9], and FACS-purified cell populations [10], as well as with a variety of insects, including fruit flies [8, 11], mosquitoes [12], aphids [13], silkworms [14], and beetles [15]. FAIRE-enriched DNA has been found to correspond to functional and developmentally relevant *cis*-regulatory elements, including transcriptional enhancers [8–10]. And when combined with a developmental time course, FAIRE-seq can successfully identify enhancers with temporal-specific activity because the timing of regulatory element accessibility often coincides with the timing of its activity. Details of the protocol are described below. The protocol is written for performing FAIRE on *Drosophila* wing imaginal discs. Guidelines and suggestions for other sample types are contained in the "**Notes**" Subheading 4.

2 Materials

2.1 Buffers and Reagents

1. Apple-agar plates: Flask 1: 200 mL apple juice, 300 mL dH_2O. Flask 2: 25 g BactoAgar, 500 mL dH_2O, stir bar. Autoclave Flask 1 and Flask 2 for 30 min. Pour contents of Flask 1 into Flask 2. Mix on magnetic stirrer until cool enough to handle. Pour into 100 × 15 mm or 60 × 15 mm plastic petri dishes. Allow to harden overnight and store at 4 °C.
2. 1 × PBS: 130 mM NaCl, 7 mM Na_2HPO_4, 3 mM NaH_2PO_4, pH to 7.4 with HCl. Autoclave to sterilize.
3. FA Fix Buffer: 4% paraformaldehyde, 50 mM HEPES pH 8.0, 100 mM NaCl, 1 mM EDTA pH 8.0, 0.5 mM EGTA pH 8.0. This buffer must be prepared fresh.
4. Quench Buffer: 1 × PBS, 125 mM Glycine, 0.01% Triton X-100.
5. Buffer A: 10 mM HEPES pH 8.0, 10 mM EDTA pH 8.0, 0.5 mM EGTA pH 8.0, 0.25% Triton X-100, 1 mM PMSF.
6. FAIRE Lysis Buffer: 2% Triton X-100, 1% SDS, 100 mM NaCl, 10 mM Tris–Cl pH 8.0, 1 mM EDTA pH 8.0.
7. TE: 10 mM Tris–Cl pH 8.0, 1 mM EDTA pH 8.0.

2.2 Equipment and Consumables

1. Screw cap tubes with 2.38 mm tungsten beads for tissue homogenization, MoBio, catalogue # 13117-50.
2. Mini-BeadBeater-8, Biospec products.
3. Branson model 450-digital sonifier.
4. QIAquick PCR purification kit (Qiagen).
5. QuBit dsDNA HS Assay kit (Invitrogen).

3 Methods

3.1 Dissection and Fixation of Drosophila Wing Imaginal Discs

1. Collect larvae of the target developmental stage. For precise developmental staging, set up an embryo-collection cage of the desired genotype. Let adults lay eggs for several hours on an apple-agar plate, then transfer the plate off the cage and incubate the plate at 25 °C. After 20–24 h, use forceps to clear all larvae from the plate, and return the plate to 25 °C. After an additional 2 h, use forceps to pick newly hatched larvae. Transfer larvae in batches of ~35 to a fresh vial containing standard *Drosophila* food. Incubate at 25 °C until the target developmental stage (*see* **Note 1**). We typically use 40 wing imaginal discs from wandering third instar larvae for a single FAIRE-seq replicate. Since each wing disc contains approximately 50,000 cells, this corresponds to about 2-million cells per replicate.

However, we have also successfully performed FAIRE-seq from 40 haltere imaginal discs, which contain approximately 10,000 cells per disc. Scale culture sizes for each tissue of interest accordingly (*see* **Note 2**).

2. Wash larvae extensively with 1 × PBS to remove any traces of food. Transfer larvae to wells containing 1 × PBS in a 9-well dissecting dish. Screen through larvae to select females (*see* **Note 3**). Dissect and invert larvae, leaving the wing discs attached to the cuticle (*see* **Note 4**). Transfer inverted larvae to cold 1 × PBS in a 9-well dissecting dish on ice.
3. Remove 1 × PBS and fix inverted larvae in 1 mL of FA Fix Buffer for 10 min at room temperature on an orbital shaker (*see* **Note 5**).
4. Stop the crosslinking reaction. Remove the FA Fix Buffer and replace with 1 mL Quench Buffer. Incubate with shaking at room temperature for 5 min. Repeat wash, for a total of 10 min of Quench incubation.
5. Remove Quench Buffer, and replace with 1 mL of ice-cold Buffer A. Keep fixed inverted larvae on ice in Buffer A until all larvae have been dissected, fixed, and quenched.
6. Dissect wing discs off cuticle in Buffer A, working in small batches to keep solution cold. Transfer discs to 1.5 mL tube on ice containing Buffer A (*see* **Note 6**).
7. Once all discs have been dissected off cuticle, pellet discs at 1500 RCF for 5 min at 4 °C. Remove Buffer A (*see* **Note 7**).

3.2 Chromatin Preparation

1. Add 1 mL of FAIRE Lysis Buffer to discs. Transfer discs to screw-cap tube containing 2.38 mm tungsten beads. Homogenize in a bead beater for six cycles of 1 min on, 2 min off at 4 °C, power setting high (*see* **Note 8**).
2. Transfer the lysate to a fresh 15 mL tube. Add FAIRE Lysis Buffer to a total volume of 1.8 mL. Sonicate to a DNA fragment size of 100 bp to 2 kb (*see* **Note 9**). We use a Branson 450-digital Sonifier and perform five 30-s cycles, with a duty cycle of 1 s on, 0.5 s off at 18% amplitude. Keep tube on ice during sonication to prevent sample heating. Let tube rest on ice for at least 2 min between sonication cycles.
3. Verify sonicated DNA fragment size by transferring 50 μL of lysate to a fresh 1.5 mL tube. Add 50 μL FAIRE Lysis Buffer and incubate the sample at 65 °C overnight to reverse crosslinks. Then add RNaseA to a final concentration of 100 μg/mL and incubate for 30 min at 37 °C. Then add Proteinase-K to a final concentration of 200 μg/mL and incubate for 2 h at 37 °C. Purify the DNA by phenol-chloroform extraction and ethanol precipitation. Resuspend precipitated DNA in 25 μL

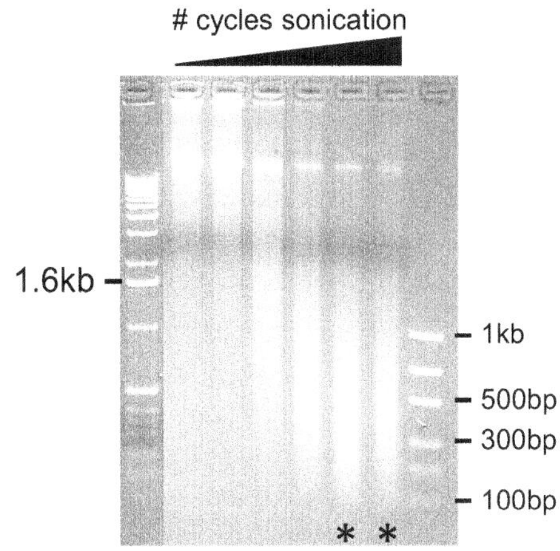

Fig. 1 The effects of increasing sonication cycle number on DNA fragment size. Agarose gel stained with ethidium bromide. Lanes 1 and 8: DNA ladders. Lanes 2–7: A fixed amount of *Drosophila* S2 cell chromatin was sonicated for an increasing number of cycles. The DNA fragment sizes in Lanes 6 and 7 (asterisks) are suitable for FAIRE-seq

TE and run on a 1.2% agarose gel. The desired fragment size is 100 bp to 2 kb, with the average size being approximately 500 bp. If the fragment size is too big, additional rounds of sonication are necessary (Fig. 1).

4. Once the desired fragment size has been achieved, pellet insoluble chromatin at 15,000 RCF for 5 min at 4 °C. Transfer the supernatant to a fresh tube, leaving behind ~30 μL to avoid carryover of DNA in the pellet. Insoluble chromatin can be discarded.

3.3 Preparation of FAIRE DNA

1. Remove a 50 μL aliquot of soluble chromatin for use as an Input sample. Reverse crosslinks, treat with RNase and Proteinase K, and purify Input DNA as described in Subheading 3.2, **step 3**. Store Input DNA at −20 °C (*see* **Note 10**).
2. Perform FAIRE on the remaining soluble chromatin lysate. Transfer 500 μL aliquots of lysate to fresh 1.5 mL tubes. Add an equal volume of phenol-chloroform to the chromatin lysate. Vortex 10-s. Centrifuge at 15,000 RCF for 5 min at 4 °C. Transfer aqueous (upper) phase to fresh 1.5 mL tube. Avoid transferring any debris that may be trapped at the interface between aqueous and organic phases.
3. Perform a second phenol-chloroform extraction. Add an equal volume of phenol-chloroform to the FAIRE DNA-containing aqueous phase. Vortex 10-s. Centrifuge at 15,000 RCF for 5 min at 4 °C. Transfer the aqueous (upper) phase to a fresh

1.5 mL tube. If the interface is not clear and free of debris, perform a third phenol-chloroform extraction.

4. Perform a chloroform extraction. Add an equal volume of chloroform/isoamyl alcohol to each tube. Vortex 10-s. Centrifuge at 15,000 RCF for 5 min at 4 °C. Transfer the aqueous (upper) phase to a fresh tube. Pool aqueous phases from samples that were divided across multiple tubes.
5. Precipitate the FAIRE-enriched DNA. Add 1/10th volume of 3 M sodium acetate (pH 5.2) and mix well. Add 10–20 μg glycogen and mix well. Add 2-volumes of 95% ethanol and mix well by inverting the capped tubes three to four times. Incubate the tubes at −80 °C for at least 30 min (*see* **Note 11**).
6. Recover tubes from −80 °C and centrifuge at 15,000 RCF for 20 min at 4 °C. Remove the supernatant without disturbing the DNA pellet. Wash the pellet with 500 μL of cold 70% ethanol. Centrifuge at 15,000 RCF for 20 min at 4 °C. Remove all traces of ethanol and let pellet dry. Resuspend FAIRE-enriched DNA pellet in 100 μL TE.
7. Heat overnight at 65 °C to reverse any remaining crosslinks. Then add 100 μg/mL RNaseA. Incubate for 30 min at 37 °C.
8. Purify DNA with QIAquick PCR purification spin column. Add 500 μL Buffer PB. Mix well by pipetting. Transfer to column. Centrifuge at 10,000 RCF for 1 min at room temperature. Empty the collection tube. Add 750 μL Buffer PE to column. Incubate for 2 min at room temperature. Centrifuge at 10,000 RCF for 1 min. Empty the collection tube. Centrifuge at 10,000 RCF for 1 min. Transfer the column to a fresh 1.5 mL tube. Centrifuge at 15,000 RCF for 1 min to remove all residual ethanol. Transfer the column to a fresh 1.5 mL tube. Add 38 μL Buffer EB directly to silica at bottom of column (*see* **Note 12**). Incubate for 1 min at room temperature. Centrifuge at 10,000 RCF for 1 min at room temperature. The eluted buffer contains the FAIRE-enriched DNA.
9. Measure DNA concentration of FAIRE-enriched DNA and Input DNA using a fluorometer (e.g., QuBit High Sensitivity dsDNA kit). The quantity of FAIRE-enriched DNA should be approximately 1–10% of the total DNA in the original sample (*see* **Note 13**).
10. FAIRE-enriched DNA and Input/genomic DNA can be stored at −20 °C indefinitely, and prepared for single-end high-throughput sequencing using a variety of commercial kits. For *Drosophila*, 5–10 million aligned reads per replicate is often sufficient for subsequent bioinformatics analysis.

4 Notes

1. When possible, take advantage of behavioral milestones or distinct morphological features to improve precision of developmental staging. For example, wandering third instar larvae can be synchronized by clearing all larvae from the walls of the vial, waiting 2 h, and then collecting newly wandering larvae for dissection. Similarly, pupal stages can be synchronized collecting animals at the white prepupal stage, which lasts only 1 h at 25 °C.
2. When possible, process experimental and control samples in parallel to avoid batch effects, which are technical in origin, from being interpreted as having a biological basis. This is preferred to processing experimental samples separately from control samples. Batch effects can arise from multiple steps in the protocol, including culture conditions, chromatin preparation, FAIRE, and high-throughput sequencing library preparation.
3. Performing FAIRE-seq only on females simplifies subsequent bioinformatics analysis by equalizing the number of sex chromosomes and autosomes.
4. Depending on how quickly the dissections can be performed, it is preferred to work in small batches to minimize the time between dissection and fixation. Fixed larvae can be left in Buffer A on ice after quenching (Subheading 3.1, **step 5**) until all larvae have been dissected.
5. These conditions are designed for *Drosophila* imaginal discs. Other conditions may be required for different tissues, developmental stages, or insect species. The objective is to achieve a light crosslinking of proteins to DNA. Therefore, as a guide, we suggest using fixation conditions that have been established for performing immunofluorescence on the target sample, with the exception of decreasing the crosslinking time by half. For example, fixation of *Drosophila* embryos for immunofluorescence is typically performed in a mixture of heptane and fixation buffer for 20 min at room temperature. For FAIRE-seq on *Drosophila* embryos, we used the same mixture of heptane and fixation buffer for 10 min at room temperature [8].
6. Since open chromatin profiles are often cell type specific, it is important to remove extraneous tissues such as trachea. Finer-detailed dissections can also be performed at this time to increase cell type specificity of FAIRE-seq data.
7. Stopping point. Dissected discs can be frozen in Liquid Nitrogen and stored at −80 °C.

8. We have noticed that bead-beating the sample improves tissue dissociation and increases reproducibility between replicates. It is possible that other means of homogenization will sufficiently dissociate the tissue prior to sonication [5].
9. It is often necessary to optimize sonication conditions for each type of sample. For example, tissues with a high amount of cuticle proteins may require additional sonication to achieve the desired DNA fragment size. Similarly, different ages, models, or brands of sonicator may vary in their performance. We suggest performing pilot experiments to empirically determine the optimal sonication conditions for a given sample, as well as the reproducibility of these results.
10. As an alternative to Input DNA, genomic DNA can be prepared from each genotype used in the experiment and sonicated to a fragment size of 100 bp to 2 kb. Sequencing Input or genomic DNA sample is essential for interpreting FAIRE-seq data because it controls for genomic features such as repetitive DNA and copy number variations.
11. Stopping point. DNA/ethanol can be stored at −80 °C indefinitely.
12. Water can be used instead of Buffer EB to elute the DNA from the column. Using water allows for the DNA to be concentrated in a speed-vac if necessary for high-throughput sequencing library preparation.
13. FAIRE-enriched DNA that is substantially greater than 10% may be indicative of samples that have been under-crosslinked. Suboptimal crosslinking of proteins to DNA may result in experimental failure due to poor signal to noise. To quantify FAIRE enrichment, use the Input sample to calculate the amount of DNA in the original sample and compare this amount to the quantity of DNA recovered in the FAIRE sample. For example, if 50 μL of soluble chromatin lysate was used for the Input sample, and 500 μL of soluble chromatin lysate was used for the FAIRE sample, the amount of DNA in the Input sample represents 1/10th of the total DNA that was present in the FAIRE sample. If the Input sample yielded 10 μg of DNA, and the FAIRE sample yielded 1 μg of DNA, then the FAIRE enrichment would be 1% of the total DNA in the original sample.

References

1. Giresi PG et al (2007) FAIRE (Formaldehyde-Assisted Isolation of Regulatory Elements) isolates active regulatory elements from human chromatin. Genome Res 17(6):877–885
2. Giresi PG, Lieb JD (2009) Isolation of active regulatory elements from eukaryotic chromatin using FAIRE (Formaldehyde Assisted Isolation of Regulatory Elements). Methods 48 (3):233–239

3. Gaulton KJ et al (2010) A map of open chromatin in human pancreatic islets. Nat Genet 42 (3):255–259
4. Song L et al (2011) Open chromatin defined by DNaseI and FAIRE identifies regulatory elements that shape cell-type identity. Genome Res 21(10):1757–1767
5. Simon JM et al (2012) Using formaldehyde-assisted isolation of regulatory elements (FAIRE) to isolate active regulatory DNA. Nat Protoc 7(2):256–267
6. Simon JM et al (2013) A detailed protocol for formaldehyde-assisted isolation of regulatory elements (FAIRE). Curr Protoc Mol Biol Chapter 21:Unit21.26
7. Mueller B et al (2017) Widespread changes in nucleosome accessibility without changes in nucleosome occupancy during a rapid transcriptional induction. Genes Dev 31 (5):451–462
8. McKay DJ, Lieb JD (2013) A common set of DNA regulatory elements shapes Drosophila appendages. Dev Cell 27(3):306–318
9. Uyehara CM et al (2017) Hormone-dependent control of developmental timing through regulation of chromatin accessibility. Genes Dev 31(9):862–875
10. Pearson JC et al (2016) Chromatin profiling of Drosophila CNS subpopulations identifies active transcriptional enhancers. Development 143(20):3723–3732
11. Davie K et al (2015) Discovery of transcription factors and regulatory regions driving in vivo tumor development by ATAC-seq and FAIRE-seq open chromatin profiling. PLoS Genet 11 (2):e1004994
12. Behura SK et al (2016) High-throughput cis--regulatory element discovery in the vector mosquito Aedes aegypti. BMC Genomics 17:341
13. Richard G et al (2017) Dosage compensation and sex-specific epigenetic landscape of the X chromosome in the pea aphid. Epigenetics Chromatin 10:30
14. Zhang Q et al (2017) Genome-wide open chromatin regions and their effects on the regulation of silk protein genes in Bombyx mori. Sci Rep 7(1):12919
15. Lai YT, Deem KD, Borràs-Castells F, Sambrani N, Rudolf H, Suryamohan K, El-Sherif E, Halfon MS, McKay DJ, Tomoyasu Y (2018) Enhancer identification and activity evaluation in the red flour beetle, *Tribolium castaneum*. Development 145(7). https://doi.org/10.1242/dev.160663 PMID: 29540499

Chapter 9

Using RAMPAGE to Identify and Annotate Promoters in Insect Genomes

R. Taylor Raborn and Volker P. Brendel

Abstract

Application of Transcription Start Site (TSS) profiling technologies, coupled with large-scale next-generation sequencing (NGS) has yielded valuable insights into the location, structure, and activity of promoters across diverse metazoan model systems. In insects, TSS profiling has been used to characterize the promoter architecture of *Drosophila melanogaster* (Hoskins et al., Genome Res 21(2):182–192, 2011) and subsequently was employed to reveal widespread transposon-driven alternative promoter usage in the fruit fly (Batut et al., Genome Res 23:169–180, 2012).

In this chapter we discuss the computational analysis of the experimental data derived from one of TSS profiling methods, RAMPAGE (RNA Annotation and Mapping of Promoters for Analysis of Gene Expression) that can be used for the precise, quantitative identification of promoters in insect genomes. We demonstrate this using the software tools GoRAMPAGE (Brendel and Raborn, GoRAMPAGE—A workflow for promoter detection by 5′-read mapping. https://github.com/BrendelGroup/GoRAMPAGE, 2016) and TSRchitect (Raborn and Brendel, TSRchitect: promoter identification from large-scale TSS profiling data. R Bioconductor package version 1.8.0 [Online]. Available: http://bioconductor.org/packages/release/bioc/html/TSRchitect.html, 2017), providing detailed instructions with the aim of taking the user from raw reads to processed results.

Key words *cis*-regulatory regions, Promoter architecture, Transcription initiation, Transcription start sites (TSSs)

1 Introduction

1.1 TSS Profiling Identifies Promoters at Genome-Scale

The promoter, which is defined in eukaryotes as the genomic region bound by RNA Polymerase II immediately prior to transcription initiation [1], is the primary locus of the regulation of gene expression. The identification of promoter regions is necessary for understanding the *cis*-regulatory signals controlling gene expression in an organism, and is also important for genome annotation. However, despite the rapid accumulation of genome sequences across metazoan and arthropod diversity, accurate annotation of promoter regions remains sparse. This is because—absent empirically-defined information—precisely identifying sequence

Susan J. Brown and Michael E. Pfrender (eds.), *Insect Genomics: Methods and Protocols*, Methods in Molecular Biology, vol. 1858, https://doi.org/10.1007/978-1-4939-8775-7_9, © Springer Science+Business Media, LLC, part of Springer Nature 2019

motifs that demarcate the promoter is unreliable. In contrast with current in silico approaches, direct mapping of TSSs identifies the location of the core promoter. Cap Analysis of Gene Expression (CAGE) [2], one of the first methods devised to identify′-ends of mRNAs at large-scale, involves selective capture of 5′-capped transcripts, first-strand reverse-transcription and ligation of a short oligonucleotide (CAGE tag).

CAGE was initially utilized by the FANTOM (Functional Annotation of the Mammalian Genome) consortium to identify promoter architecture in human and mouse [3], providing the first glimpse of the global landscape of transcription initiation. At the onset of the next-generation sequencing (NGS) era, CAGE was coupled with massively-parallel sequencing to define 5′-mRNA ends at large scale. This advance provided more extensive coverage of the expressed transcriptome and provided increased sensitivity for quantitative measurements of promoter activity.

1.2 Promoter Architecture of Drosophila melanogaster

Hoskins and colleagues [4] performed CAGE in *D. melanogaster* as part of the modENCODE consortium, identifying promoters at large-scale and characterizing the promoter architecture of an insect genome for the first time with TSS profiling. The authors found that TSS distributions at *Drosophila* promoters exhibit a range of shapes that can be generally grouped into two major classes: *peaked* and *broad*. This confirmed the original finding of Rach and colleagues [5], which was done using publicly-available expressed sequence tags (ESTs). Peaked promoters tend to have a single, major TSS position occupying a narrow genomic region, whereas broad promoters lack a single, major TSS and contain TSSs across a wider region [5, 6]. The authors also showed a strong association between promoter class and motif composition (consistent with previous findings [5, 7]). Peaked promoters were associated with positionally-enriched *cis*-regulatory motifs including TATA, Initiator (Inr), and DPE (Downstream Promoter Element), while broad promoters contained an enrichment of less well-characterized motifs, including *Ohler6* and *Ohler7* [8]. The existence of at least two promoter classes appears to be conserved among metazoans and has been reported (using TSS profiling methods) in insects, cladocerans [9], fish [10], and mammals [6, 11].

1.3 Promoter Structure of Insects

Beyond *D. melanogaster*, few investigations have utilized TSS profiling in insect genomes. As a consequence, what is known about promoter architecture in insects is largely restricted to the *Drosophila* genus. As part of the modENCODE effort, CAGE was performed in multiple tissues and developmental stages of the *Drosophila pseudoobscura*. TSSs were found to be highly similar between species: 81% of TSSs of aligned, CAGE-identified TSSs from *D. pseudoobscura* were positioned within 20nt of their counterparts in *D. melanogaster*. An enrichment of the CA dinucleotide was detected at the TSS ([−1, +1]), and the motifs corresponding to TATA, Inr, and DPE were positioned at the same locations relative to the TSS in both species.

The only other insect species for which TSS profiling has been applied is the Tsetse fly (*Glossina morsitans morsitans*) [12]. Using TSS-seq (specifically Oligo-capping; for details see [13]), the authors identified 3134 promoters associated with 1424 genes. The authors found a preference for CA and AA dinucleotides at the TSSs and observe the major core promoter elements observed in *Drosophila*: TATA, Inr, DPE, in addition to MTE (the Motif Ten Element). As in *D. melanogaster*, peaked promoters were more likely to contain TATA and Inr than broad promoters. While the taxonomic sampling of species for TSS profiling has been limited, the existing studies are sufficient to provide a general picture of insect promoter architecture. A major demarcation between the promoter architecture of insects and mammals appears to be the large fraction of mammalian promoters found in CpG islands [12]. CpG island promoters (CPIs) form the largest class of promoter in mammals [14]; by contrast, CPIs are not known to exist as a class in invertebrates.

1.4 Paired-End TSS Profiling with RAMPAGE

A notable recent methodological advance in TSS Profiling is RAMPAGE [15, 16], a protocol for 5′-cDNA sequencing that combines cap trapping and template-switching with paired-end sequence information (*see* Fig. 1). As with CAGE and other TSS profiling methods, RAMPAGE reads are aligned, to obtain TSSs and clustered to identify Transcription Start Regions (TSRs), which are enrichments of TSSs consistent with promoters (Fig. 2a). A key advantage of generating paired-end sequence is transcript connectivity, which provides a direct link between a given 5′-end and its associated mRNA molecule [15] (Fig. 2b). Because short or spurious RNAs are found within the transcriptome, transcript connectivity allows the TSSs (and thus promoters) of full-length mRNAs to be unambiguously identified, which benefits genome annotation and improves interpretation of transcript species. There are other TSS profiling methodologies, provide paired-end information, although these methods differ (with each other and with RAMPAGE) in the ways capped RNA is captured and processed into finished libraries. These include PEAT (Paired-end analysis of transcription) [7] and nanoCAGE [17, 18]. PEAT has been applied in two species to date: *D. melanogaster* [7] and the model plant *Arabidopsis thaliana* [19], whereas nanoCAGE has been applied to mammalian systems. While this chapter will discuss the processing and analysis of RAMPAGE libraries, the code and tools we present here are capable of handling any other TSS profiling read datasets.

Batut and colleagues [15] generated libraries from total RNA isolated from 36 stages across the life cycle of *D. melanogaster*, generating a comprehensive gene expression and promoter atlas

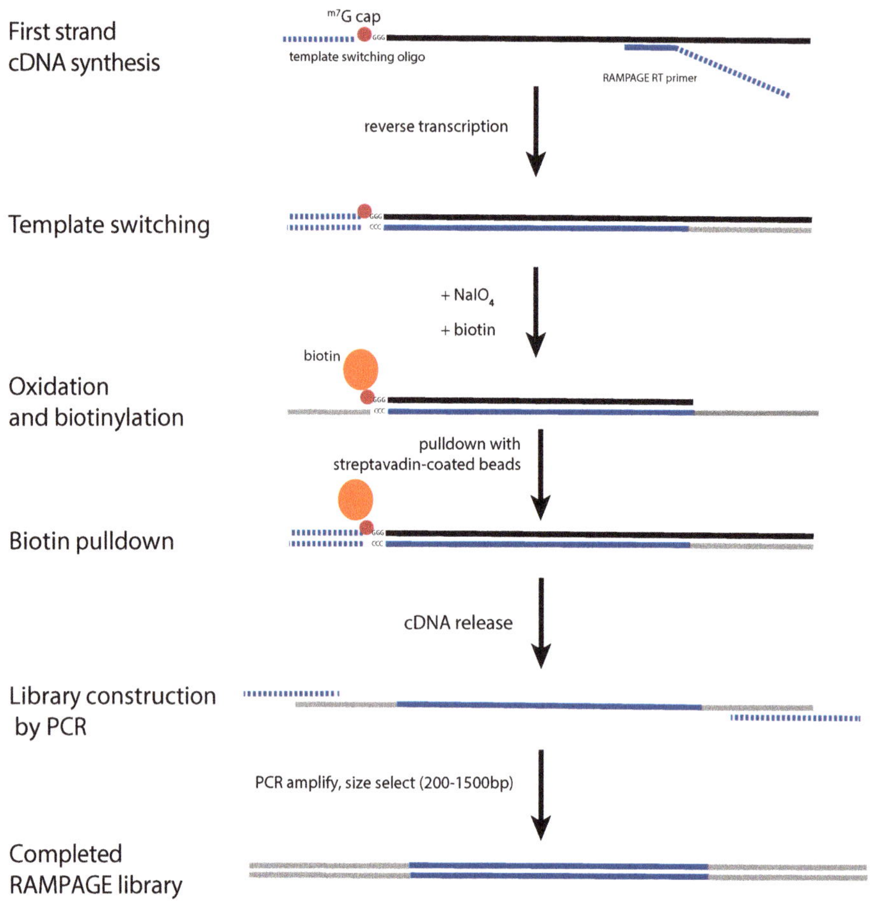

Fig. 1 A brief summary of the RAMPAGE protocol. Starting with high-quality total RNA, first-strand cDNA synthesis is initiated using a cap-bound oligonucleotide and a custom RAMPAGE RT primer, creating a double-stranded DNA-RNA hybrid molecule. Next, the 5′-m7G cap is oxidized, bound with biotin and pulled down with streptavidin-coated beads. The single-stranded cDNA molecules are released and the final RAMPAGE library construction is completed with PCR using custom oligonucleotides, followed by size-selection. This illustration was adapted from [16]

for fruit fly and demonstrating the utility of RAMPAGE. RAMPAGE is currently being applied as part of the latest iteration of ENCODE [20] to identify promoters in diverse human tissues [21], but as of this writing it has not been applied to any non-*Drosophila* insect model system.

In anticipation of the future application of TSS profiling toward other insect model systems, we discuss in this chapter a well-documented protocol for the computational processing and analysis of RAMPAGE data, using selected libraries from Batut et al. [15]. This method consists of two parts: first, we discuss how to process, filter, and align the sequenced RAMPAGE libraries to the *D. melanogaster* genome. Second, we show how to identify TSSs and promoters from the aligned sequences and associate them with

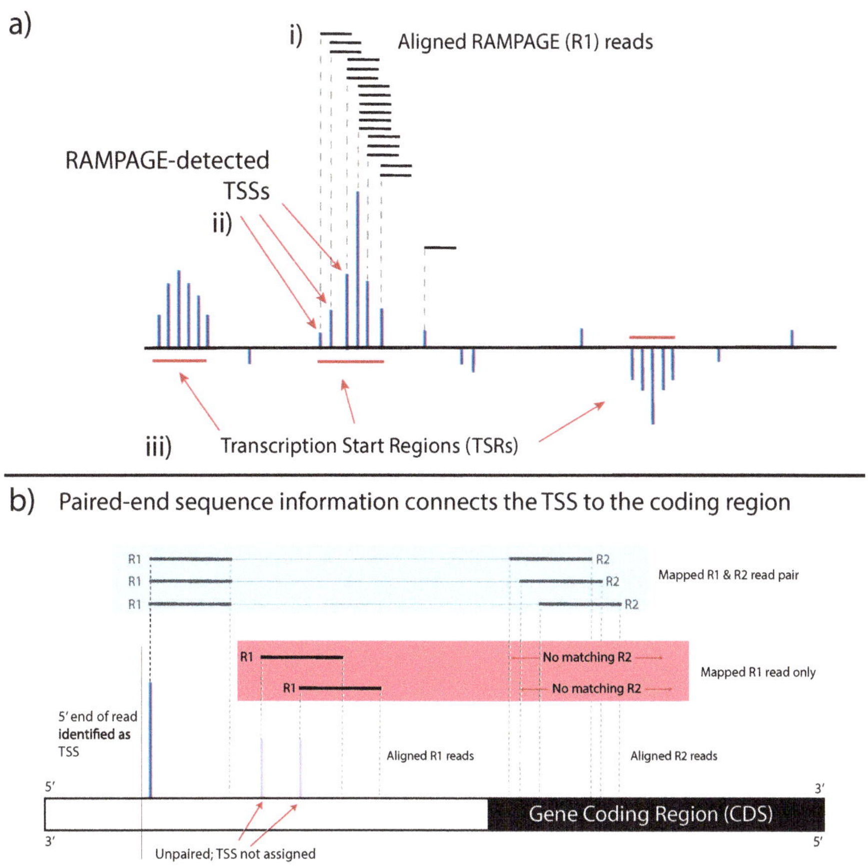

Fig. 2 An overview of promoter identification using RAMPAGE. (**a**) RAMPAGE reads are aligned to the genome. The 5′-most genomic coordinate from each properly-paired R1 read is estimated as a TSS. The abundance of mapped 5′-ends at a given TSS is a measure of its abundance. TSSs above a minimum threshold will be clustered into TSRs. (**b**) RAMPAGE-derived paired-end sequence information provides a connection between a 5′-mRNA end and a gene coding region. Only properly-paired R1 reads (i.e., with an aligned R2 read) are identified as TSSs and then included in the downstream clustering procedure described in part (**a**)

coding regions. In closing, we will consider further applications of this data and discuss the utility of reproducible workflows in bioinformatic analysis.

2 Materials

The example analyses described herein require a workstation capable of doing modern bioinformatics; minimally a reasonably-appointed laptop. An intermediate understanding of the Linux/Unix command line will be extremely useful, although we make efforts to explain the procedures with clarity. In addition, it will likely be necessary for the participant to have superuser privileges on the machine. If you do not have a machine (or lack access to one) that meets these requirements, it is recommended that you consider cloud-based cyberinfrastructure, including Amazon Web Services (AWS; https://aws.amazon.com/), CyVerse (http://www.cyverse.org/) [22] or JetStream (https://jetstream-cloud.org/) [23]. The former is a well-known pay-per-use solution, while the latter two are NSF-funded resources that make compute allocations freely available to the public.

For many users, the cyberinfrastructure approach is a convenient solution, particularly when providers offer task-dedicated virtual machines. In that case, the user essentially rents a fully equipped computer with all necessary software pre-installed and sufficient resources for the intended job. For the computational workflows discussed here, researchers can check out an instance of the "bgRAMOSE" image at JetStream which comes with all Brendel Group software (https://brendelgroup.github.io/) as well as other useful bioinformatics tools enabled. For details on JetStream and for instructions on how to check out a JetStream image, please read documentation for JetStream found at https://iujetstream.atlassian.net/wiki/spaces/JWT/overview.

2.1 Hardware

1. x86-64 compatible processors
2. 16 GB RAM
3. 80 GB+hard disk space

2.2 Operating System

- 64 bit Linux (preferred) or Mac OS X (with Command Line Tools from XCode)

2.3 Software

Below is a list of the software packages required for this demonstration (*see* **Note 1**).

Sequence Retrieval

1. SRA Toolkit [24] (https://www.ncbi.nlm.nih.gov/sra/docs/toolkitsoft/)

GoRAMPAGE

1. GoRAMPAGE [25] (https://github.com/BrendelGroup/GoRAMPAGE)
2. fastq-multx [26] (https://github.com/brwnj/fastq-multx)
3. FASTX-Toolkit [27] (http://hannonlab.cshl.edu/fastx_toolkit/Index.html)
4. TagDust2 [28] (https://sourceforge.net/projects/tagdust/)
5. Samtools [29] (http://www.htslib.org/doc/samtools.html)
6. STAR [30] (https://github.com/alexdobin/STAR)

TSRchitect

1. R (v. 3.4 and up) [31] (https://www.r-project.org/)
2. Bioconductor (v. 3.5 and up) [32] (http://bioconductor.org/)

3. *TSRchitect* [33] (http://bioconductor.org/packages/release/bioc/html/TSRchitect.html)
4. Various R package dependencies (*see* Subheading 3)

2.4 Demonstration

We created an online demonstration (demo) to serve as a companion to this chapter, which contains both scripts and select files to assist you in completing this tutorial. Please find the repository at here (in the folder "demo/MMB/"): https://github.com/BrendelGroup/GoRAMPAGE (*see* **Note 2**).

2.5 Installation of R Packages

For installation of the software listed above, please follow the instructions provided by each respective package. Part of our analysis will require the use of R packages found in the Bioconductor suite [32] (*see* **Note 3**). To install Bioconductor, please type the following from an R console:

```
source("https://bioconductor.org/biocLite.R")
biocLite()
```

We will use the R package *TSRchitect* to identify promoters from aligned RAMPAGE libraries. Prior to running the analysis, it will be necessary to install a series of prerequisite packages to *TSRchitect* from Bioconductor. Please install these packages, followed by *TSRchitect* (as before, from an R console):

```
source("https://bioconductor.org/biocLite.R")
biocLite(c("AnnotationHub", "BiocGenerics", "BiocParallel",
 "ENCODExplorer",  "GenomicAlignments", "GenomeInfoDb",
 "GenomicRanges", "IRanges", "Rsamtools",
 "rtracklayer", "S4Vectors",
 "SummarizedExperiment"))

biocLite("TSRchitect")
```

Finally, please confirm that *TSRchitect* has been installed correctly by loading it from your R console as follows:

```
library(TSRchitect) #loading TSRchitect
```

3 Methods

3.1 Retrieving the RAMPAGE Sequence Data from NCBI

To begin our analysis, we must download the RAMPAGE data to our workstation. We will utilize tools provided by the SRA Toolkit, which should already be installed on your machine (*see* Subheading 2). The command *fastq-dump* allows one to directly retrieve data from the GEO database using the appropriate identifier(s). While there are 36 RAMPAGE libraries in the Batut et al. paper, we will select a subset of these to analyze here. We will compare samples from selected embryonic (E01h–E03h) and larval (L1–L3) tissues, representing the beginning and end of embryonic development. For more information about the experiment and the available RAMPAGE libraries, please see the following link: https://www.ncbi.nlm.nih.gov/Traces/study/?acc=SRP011193.

First, let's proceed with downloading the libraries from early embryonic tissues (*see* **Note 4**). We will make a new folder (entitled "`fastq_files/`") to house these files.

```
mkdir fastq_files
cd fastq_files


fastq-dump --split-files SRR424683
fastq-dump --split-files SRR424684
fastq-dump --split-files SRR424685
```

We continue by downloading the data from late larval tissues.

```
fastq-dump --split-files SRR424707
fastq-dump --split-files SRR424708
fastq-dump --split-files SRR424709
```

Once the download of the aforementioned files are complete, you should see a total of 12 (6 × 2) separate FASTQ files in your current working directory:

```
ls -1 *.fastq | wc -l
```

3.2 Creating Symlinks to the Files

Our workflow expects fastq files that have the format "*.R1/R2.clipped.fq". Rather than rename them, we can simply create brand new symbolic links (symlinks) to the files, as follows:

```
cd ..
mkdir -p output/reads/clipped
cd output/reads/clipped

#embryonic libraries
ln -s ../../../fastq_files/SRR424683_1.fastq E01h.R1.clipped.fq
ln -s ../../../fastq_files/SRR424683_2.fastq E01h.R2.clipped.fq
ln -s ../../../fastq_files/SRR424684_1.fastq E02h.R1.clipped.fq
ln -s ../../../fastq_files/SRR424684_2.fastq E02h.R2.clipped.fq
ln -s ../../../fastq_files/SRR424685_1.fastq E03h.R1.clipped.fq
ln -s ../../../fastq_files/SRR424685_2.fastq E03h.R2.clipped.
fq#larval libraries
ln -s ../../../fastq_files/SRR424707_1.fastq L1.R1.clipped.fq
ln -s ../../../fastq_files/SRR424707_2.fastq L1.R2.clipped.fq
ln -s ../../../fastq_files/SRR424708_1.fastq L2.R1.clipped.fq
ln -s ../../../fastq_files/SRR424708_2.fastq L2.R2.clipped.fq
```

```
ln -s ../../../fastq_files/SRR424709_1.fastq L3.R1.clipped.fq
ln -s ../../../fastq_files/SRR424709_2.fastq L3.R2.clipped.fq

cd ../../.. #returning to the output directory
```

3.3 Downloading Genomic Data from D. melanogaster

Now that we have the fastq files from the RAMPAGE libraries downloaded and named appropriately, we now must retrieve the genome assembly and rRNA sequences from *D. melanogaster*. The genome assembly is required for aligning the RAMPAGE reads, and the rRNA sequences are required to filter out matching reads in the sequenced RAMPAGE libraries. Because our sample is intended to contain only capped RNAs, any rRNA sequences we observe in these RAMPAGE libraries are contaminants that must be removed.

Please make note of the rRNA sequences, found in the file "`Dmel_rRNA.fasta`", from the folder `additional_files` folder in the demo (*see* **Note 5**).

We will then download a version of the *D. melanogaster* genome assembly from ENSEMBL (www.ensembl.org) [34]. To retrieve the genome assembly, please do the following:

```
mkdir genome
cd genome
wget ftp://ftp.ensembl.org/pub/release-78/fasta/
drosophila_melanogaster/dna/Drosophila_melanogaster.BDGP5.dna.toplevel.fa.gz
#uncompressing the file
gzip -d Drosophila_melanogaster.BDGP5.dna.toplevel.fa.gz
cd ..
```

3.4 Filtering and Alignment of RAMPAGE Reads Using GoRAMPAGE

At this stage we are ready to commence with the rRNA filtering and alignment of the RAMPAGE libraries. We will use GoRAMPAGE, a tool we developed, to perform these tasks in a concerted workflow. GoRAMPAGE runs TagDust [28] to remove rRNA and low-complexity reads and STAR [30] to align RAMPAGE (or other paired-end) reads to a given genome assembly.

3.4.1 Setting Up the GoRAMPAGE Job

Please refer to the script "`GoRAMPAGE_script_MMB.sh`" and (using a text editor) provide the appropriate paths to the genome assembly, output directory (see above), and rRNA sequences (*see* **Note 6**). GoRAMPAGE jobs can optionally be run in parallel (*see* **Note 7**). The script can be executed as follows:

```
#vi GoRAMPAGE_script_MMB.sh #updating with a text editor
./GoRAMPAGE_script_MMB.sh
```

If everything is working correctly, you should start to see the results of the job being written to the file "errScript." You can inspect the progress during the run using the *less* command.

```
less -S errScript
```

Should the run fail before completion, any associated error messages will be printed to the errScript file. Once the job is complete, you should see the message "GoRAMPAGE job is complete!" appear on the command-line terminal.

3.4.2 Inspecting the rRNA Filtering Results

To evaluate the results from Step 3 (rRNA filtering), please navigate to the top level of the "output" directory and open the file "LOGFILES". You'll see the recorded progress of the program Tagdust and a record of the results. We notice that (for the L3h library) 1046448 of reads (78.1%) were "extracted," meaning that slightly more than 20% of reads were removed because of matches with ribosomal sequences. The removed reads from all libraries are found in the `"dusted_discard"` directory, and the extracted reads are found in the current directory. Due to their sheer abundance within cells, ribosomal RNA sequences are an inevitable contaminant within TSS profiling libraries. For the purpose of our analysis, it is important that these sequences be removed, which is what has been completed here.

Since this step was conducted appropriately, we can proceed to the next step.

3.4.3 Evaluating the Alignments

The folder `"alignments/"` in your GoRAMPAGE output folder will now contain 6 .bam files, each representing the distinct RAMPAGE libraries selected for our analysis. Typing "ls -l" from the command line will show that these files are symlinks to the original alignment files found in the `"STARoutput/"` directory. "STARoutput/", as its name suggests, contains the output from the STAR alignment, and this includes the alignment files "*.sortedByCoord.out.bam", and four additional log files. The files with the suffix "*.STAR.Log.final.out" each contain a summary of the alignment, such as the number of input reads, the percentage of uniquely-mapped reads, and the percentage of unmapped reads. An inspection of these log files indicates that the alignments have similar mapping rates (~70–80%), a reasonable outcome for our purposes.

Now that our RAMPAGE libraries are filtered and aligned, we can commence with the second half of our analysis.

3.5 Promoter Identification from Aligned RAMPAGE Libraries

We can now use the prepared alignment files to identify TSSs and promoters from the selected RAMPAGE libraries. There are currently several tools available for this purpose. *CAGEr*, developed by Haberle [35], was utilized to perform TSS identification as part of the FANTOM5 efforts. We will use *TSRchitect* in this demonstration, since it was specifically designed to analyze paired-end TSS profiling datasets, and also because it is more flexible with respect to model system (i.e., it does not require a corresponding *BSGenome* [36] package). The latter feature will be helpful when analyzing the non-*D. melanogaster* TSS profiling datasets that we expect to be generated in the near future.

3.5.1 Setting Up the Analysis

TSRchitect, the package we'll use for this analysis, is an R package available in the Bioconductor suite of genomics tools [32]. It makes use of existing packages and data structures within this environment, where available, to identify promoters from sequence alignments. Since you have already installed *TSRchitect* and its dependencies (*see* Subheading 2.3), we are set to proceed.

There are two general ways one can choose to run *TSRchitect*. The first is interactively, i.e., typing the instructions directly into an R console. While this is a perfectly acceptable way to run analyses using package, for larger jobs it will likely be more efficient (and likely more reproducible) to run a dedicated R script. We have provided sample scripts to make it easier for you to set up an R script. The two scripts are identical with a lone exception: one is set up to run in parallel (`"TSRchitect_parallel_MMB.R"`), while the other is written to run in serial (`"TSRchitect_serial_MMB.R"`). Please select the script that best suits your computing resources. In the section to follow, we will go through the output of the analysis. For further details on how to use *TSRchitect*, please see its documentation at its Bioconductor page found here: https://www.bioconductor.org/packages/release/bioc/html/TSRchitect.html.

3.5.2 Running the Analysis

To run *TSRchitect* using the batch script, provide full paths for the variables "BAMDIR" and "DmAnnot" in the script provided (*see* **Note 8**). *BAMDIR* should be a path to the subdirectory "alignments/" in the RAMPAGE output directory you specified earlier, and *DmAnnot* should be a full path to the *D. melanogaster* gene annotation listed above.

Once this is complete, we can run the batch script from the Linux command-line as follows:

```
R CMD BATCH TSRchitect_parallel_MMB.R #or use 'serial script
#assumes variables BAMDIR and DmAnnot have already been set
bg #puts this job in the background
```

Once the job is underway, you can monitor its progress by looking at the contents of the .Rout file (in this case, `"TSRchitect_parallel_MMB.Rout"`).

3.5.3 Reviewing the TSRchitect Script

Before we evaluate the results (which will have been written to your working directory after running the batch script), there are some important aspects of the analysis to review. We discuss these for informational purposes only; it will not be necessary to perform these commands separate from the batch script provided. First, we must initialize the *tssObject* (which stores the information about the experiment) appropriately (*see* **Note 9**).

The inputs in this case are BAM files; *TSRchitect* also accepts input in BED format.

```
DmRAMPAGE <- loadTSSobj(experimentTitle = "RAMPAGE Tutorial", \
 inputDir=BAMDIR, isPairedBAM=TRUE, \
 sampleNames=c("E1h","E2h", "E3h", "L1", "L2", "L3"), \
 replicateIDs=c(1,1,1,2,2,2))
```

Next, we extract TSSs from the alignment files:

```
DmRAMPAGE <- inputToTSS(DmRAMPAGE)
```

A critical step in our analysis is identifying TSRs from the aligned TSS data; to do this, we use the function *determineTSR*. We have selected the job to run on 4 cores in this example (*n. cores*=4). Please enter the number of cores appropriate for your system. Because we want to identify TSRs from every one of the selected RAMPAGE libraries, we specify *tssSet*="all". The parameter *tagCountThreshold* was set to 25, meaning that only TSSs supported by 25 or more 5′ RAMPAGE reads will be included within a TSR. Setting *writeTable* to "TRUE" means that the identified TSRs from each set will be written to the working directory.

```
DmRAMPAGE <- determineTSR(experimentName=DmRAMPAGE, n.cores=4, \
 tsrSetType="replicates", tssSet="all", tagCountThreshold=25, \
 clustDist=20, writeTable=TRUE)
```

TSRchitect can incorporate the tag abundances from each of the samples and append them to the list of identified TSRs. This is useful for downstream analysis of differential expression.

```
DmRAMPAGE <- addTagCountsToTSR(experimentName=DmRAMPAGE, \
tsrSetType="replicates",  tsrSet=1, tagCountThreshold=10, \
   writeTable=TRUE)
```

We can use *TSRchitect* to import an annotation file (or, alternatively, use an existing one from *AnnotationHub*) and use it to associate our set of identified TSRs with coding genes. We can specify the maximum distances (both up- and downstream) between the TSR and the annotation using the arguments *upstreamDist* and *downstreamDist*.

```
DmRAMPAGE <-importAnnotationExternal(experimentName=DmRAMPAGE, \
 fileType="gff3",  annotFile=DmAnnot)

DmRAMPAGE <-addAnnotationToTSR(experimentName=DmRAMPAGE, \
 tsrSetType="replicates", tsrSet=1, \
upstreamDist=1000, downstreamDist=200, feature="gene", \
 featureColumnID="ID", writeTable=TRUE)
```

Now we have generated a set of identified TSSs, TSRs from all six RAMPAGE libraries, and have associated the identified TSRs with annotated genes. Next, we will merge the libraries into two samples according to condition: early embryonic (E1h, E2h, E3h) and late larval (L1, L2, L3) using the information we provided when we initialized the *tssObject* at the start of this section. After merging, we will identify promoters (1) within the merged samples and (2) within the entire dataset combined, and associate with the *D. melanogaster* gene annotation as described previously (not shown).

```
#merging the sample data into two groups
DmRAMPAGE <- mergeSampleData(DmRAMPAGE)

# ... identifying TSRs from the merged samples:
DmRAMPAGE <- determineTSR(experimentName=DmRAMPAGE, \
  n.cores=4, tsrSetType="merged", \
   tssSet="all", tagCountThreshold=40, \
   clustDist=20, writeTable=TRUE)
```

3.5.4 Evaluating the Results

Our analysis of TSS profiling data using *TSRchitect* is now complete. A snapshot of a representative sample of small set of aligned RAMPAGE libraries is shown in Fig. 3. Your working directory should now contain the following:

- TSSs from each sample, e.g. TSSset-1.txt: (6)
- TSRs from each sample (in both .txt and .tab formats): (12)
- TSRs from each merged group (in both .txt and .tab formats): e.g., TSRsetMerged-1.txt: (4)
- TSRs from the combined set of TSSs: TSRsetCombined.tab: (1)

Let's briefly review the files (*see* **Note 10**). We can quickly obtain the counts on the command line, as follows:

```
wc -l *.tab
8377 TSRset-1.tab
6159 TSRset-2.tab
4814 TSRset-3.tab
17924 TSRset-4.tab
11851 TSRset-5.tab
3242 TSRset-6.tab
13986 TSRsetCombined.tab
7344 TSRsetMerged-1.tab
12126 TSRsetMerged-2.tab
85823 total
```

We will see that we have identified between roughly 3200 and 18,000 TSRs within the individual RAMPAGE samples, which is attributable to the differences in library sizes. We detect 7344 TSRs within the early embryonic samples ("TSRsetMerged-1.tab") and 12,126 TSRs in the late larval samples ("TSRsetMerged-2.tab"). Within the combined samples ("TSRsetCombined.tab") we find

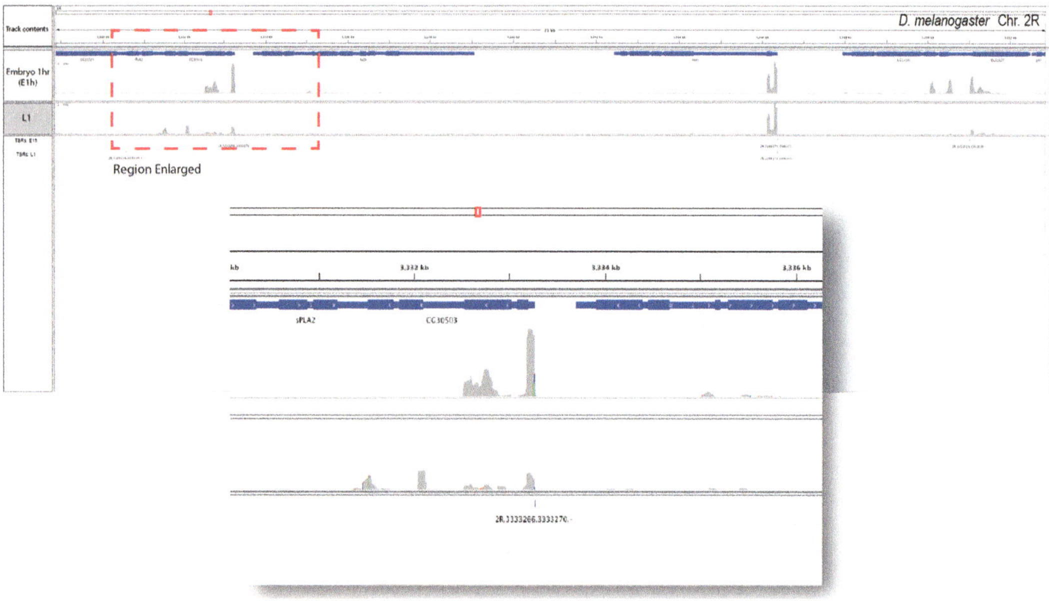

Fig. 3 An overview of the TSS profiling information provided by RAMPAGE. A representative visualization of RAMPAGE peaks (i.e., clusters of properly-aligned RAMPAGE reads) within an arbitrarily-selected genomic region of *D. melanogaster* chromosome 2R is shown, along with the corresponding gene annotation within this region. RAMPAGE data from two RAMPAGE libraries from Batut et al. [15] are shown, which were generated from RNA isolated from developmental stages E1h and L1, *see* Subheading 3. For each library, the abundance of RAMPAGE reads that align to a given site within the genome is represented by density plots (shown in gray). Gene models are shown in blue, where the thickened line represents exons and thin lines represent introns. The locations of TSRs identified by *TSRchitect* are shown in the two tracks from the bottom of the image. A single region, highlighted with the red dashed line is enlarged (the *Inset*) to show further detail of a selected gene and RAMAPGE signals. In some cases, the abundance of signal at 5′-ends is roughly equivalent, whereas in others the observed signal is substantially higher (*see* Inset). The original images are screenshots generated in the Integrated Genomic Viewer (IGV; http:/software.broadinstitute.org/software/igv/) [37]. Where necessary, additional annotation was added using Adobe Illustrator

13,986 TSRs, which is similar to the number reported by Hoskins et al. [4].

In addition to identifying the position of a given TSRs, *TSRchitect* records other useful information about its properties. The *width* of a TSR refers the span of the genomic region it occupies (in bp), and the *Shape Index* (SI) is measure of the relative peakedness of the TSR. We can see an example of this in the file "TSRsetMerged-1.txt" (*see* **Note 11**).

seq	start	end	strand	nTAGS	tsrWdth	tsrSI	featureID
2L	67043	67044	+	270	2	1	NA
2L	74089	74115	+	341	27	0.13	NA
2L	94739	94752	+	1650	14	0.55	FBgn0031216
2L	102386	102386	+	284	1	2	FBgn0031217

3.6 Summary

The workflow provided here is intended to serve as a useful entry point for the analysis of TSS profiling data in insects. On the computational side, we have provided a set of open source tools so that the uninitiated genome scientist can begin to analyze RAMPAGE (and other forms of TSS profiling data) quickly. While the analysis centered on *D. melanogaster* via the use of public datasets, it is anticipated that this will assist groups who may be interested in performing TSS profiling in their preferred insect model system. The application of TSS profiling technology across a more representative swathe of insect diversity will improve our understanding of the positions and general structure *cis*-regulatory regions in this phylum.

4 Notes

1. Please consult the GoRAMPAGE documentation found here:

   https://github.com/BrendelGroup/GoRAMPAGE.

   Installation instructions for the prerequisites of GoRAMPAGE (which includes some of the items listed) are found at the following link:

   https://github.com/BrendelGroup/GoRAMPAGE/tree/master/src.

2. On Linux, the installation of a few packages is necessary in order to install Bioconductor packages using *biocLite()*.

   To install them using Ubuntu:

```
apt-get install libssl-dev
apt-get install libcurl4-openssl-dev
apt-get install libxml2-dev
```

   If you do not have Ubuntu, use the commands necessary to install the above packages on your Linux distribution.

3. You can clone the entire GoRAMPAGE repository (which includes the contents of the demo) to your workspace on the command line using git, as follows:

```
https://github.com/BrendelGroup/GoRAMPAGE
cd demo/MMB
```

   The `"scripts/"` folder in the demo contains code for you to run the two major workflows described in this chapter. The `"additional_files/"` folder contains the following files which are necessary for the analysis: (1) a fasta file containing ribosomal RNA sequences for *D. melanogaster* (`Dmel_rRNA.fasta`) and (2) a gene annotation for *D. melanogaster* (`Drosophila_melanogaster.BDGP5.78.gff`).

4. Since these fastq files are paired-end, we use the argument *–split-files* to generate separate files for each read pair.
5. If you are running this on a cluster with a job scheduler, you'll need to add the necessary headers to the top of the script and submit the job in the appropriate manner.
6. The rRNA sequences were retrieved separately from Genbank at NCBI [38].
7. For parallel execution, GoRAMPAGE uses the Linux package *GNU parallel* [39]. Please see the GoRAMPAGE documentation for more information.
8. To do this, please edit the batch script `TSRchitect_serial_MMB.R` with a text editor of your choice.
9. Because the samples provided derive from related developmental stages, we will merge them for annotation purposes using the argument *replicateIDs* (though it must be emphasized that they are not replicates).
10. All of *TSRchitect*'s output files are labeled according to the order that they are loaded onto the *tssObject*. For example, *TSSset-1.txt* corresponds to the first RAMPAGE dataset (in our case E1h), and *TSSset-2.txt* corresponds to the second RAMPAGE dataset (for this example E2h), and so on. You can check which datasets are loaded on the *tssObject* by simply entering its name on an R console. Please see the *TSRchitect* documentation for more information.
11. Please note that only selected columns are shown; *TSRchitect* v 1.8 and above provides more columns (with additional TSR-related metrics) than is displayed here.

Acknowledgements

The authors would like to thank Philippe Batut for generous technical assistance with the RAMPAGE protocol, and to Nathan Keith for his help establishing the protocol in our laboratory. The authors are grateful to Thomas W. McCarthy for his help testing the code and providing editorial feedback.

Disclosure Declaration The authors declare that they have no competing interests.

References

1. Kadonaga JT (2012) Perspectives on the RNA polymerase II core promoter. Wiley interdisciplinary reviews: developmental biology, vol 1 (1). Wiley, New York, pp 40–51
2. Kodzius R, Kojima M, Nishiyori H, Nakamura M, Fukuda S, Tagami M et al (2006) CAGE: cap analysis of gene expression. Nat Methods 3(3):211–222

3. Carninci P, Kasukawa T, Katayama S, Gough J, Frith MC, Maeda N et al (2005) The transcriptional landscape of the mammalian genome. Science (New York, NY) 309 (5740):1559–1563
4. Hoskins RA, Hoskins RA, Landolin JM, Landolin JM, Brown JB, Brown JB et al (2011) Genome-wide analysis of promoter architecture in Drosophila melanogaster. Genome Res 21(2):182–192
5. Rach EA, Yuan H-Y, Majoros WH, Tomancak P, Ohler U (2009) Motif composition, conservation and condition-specificity of single and alternative transcription start sites in the Drosophila genome. Genome Biol 10(7): R73
6. Lenhard B, Sandelin A, Carninci P (2012) Metazoan promoters: emerging characteristics and insights into transcriptional regulation. Nat Rev Genet 13(4):233–245
7. Ni T, Corcoran DL, Rach EA, Song S, Spana EP, Gao Y et al (2010) A paired-end sequencing strategy to map the complex landscape of transcription initiation. Nat Methods 7 (7):521–527
8. Ohler U, Liao G-c, Niemann H, Rubin GM (2002) Computational analysis of core promoters in the Drosophila genome. Genome Biol 3 (12):research0087.1–0087.12
9. Raborn RT, Spitze K, Brendel VP, Lynch M (2016) Promoter architecture and sex-specific gene expression in Daphnia pulex. Genetics 204(2):593–612
10. Nepal C, Hadzhiev Y, Previti C, Haberle V, Li N, Takahashi H et al (2013) Dynamic regulation of the transcription initiation landscape at single nucleotide resolution during vertebrate embryogenesis. Genome Res 23 (11):1938–1950
11. Carninci P, Sandelin A, Lenhard B, Katayama S, Shimokawa K, Ponjavic J et al (2006) Genome-wide analysis of mammalian promoter architecture and evolution. Nat Gen 38(6):626–635
12. Mwangi S, Attardo G, Suzuki Y, Aksoy S, Christoffels A (2015) TSS seq based core promoter architecture in blood feeding Tsetse fly (Glossina morsitans morsitans) vector of Trypanosomiasis. BMC Genomics 16(1):722
13. Tsuchihara K, Suzuki Y, Wakaguri H, Irie T, Tanimoto K, Hashimoto S-i et al (2009) Massive transcriptional start site analysis of human genes in hypoxia cells. Nucleic Acids Res 37 (7):2249–2263
14. Cvetesic N, Lenhard B (2017) Core promoters across the genome. Nat Biotechnol 35 (2):123–124
15. Batut PJ, Dobin A, Plessy C, Carninci P, Gingeras TR (2012) High-fidelity promoter profiling reveals widespread alternative promoter usage and transposon-driven developmental gene expression. Genome Res 23:169–180
16. Batut PJ, Gingeras TR (2013) RAMPAGE: promoter activity profiling by paired-end sequencing of 5'-complete cDNAs. In: Ausubel FM *et al* (eds) Current protocols in molecular biology. Wiley, Hoboken, pp 25B.11.1–25B.11.16
17. Plessy C, Bertin N, Takahashi H, Simone R, Salimullah M, Lassmann T et al (2010) Linking promoters to functional transcripts in small samples with nanoCAGE and CAGEscan. Nat Methods 7(7):528–534
18. Cumbie JS, Ivanchenko MG, Megraw M (2015) NanoCAGE-XL and CapFilter: an approach to genome wide identification of high confidence transcription start sites. BMC Genomics 16(1):528
19. Morton T, Petricka J, Corcoran DL, Li S, Winter CM, Carda A et al (2014) Paired-end analysis of transcription start sites in Arabidopsis reveals plant-specific promoter signatures. Plant cell 26(7):2746–2760 (2014)
20. ENCODE Project Consortium (2012) An integrated encyclopedia of DNA elements in the human genome. Nature 489(7414):57–74
21. Consortium E (2017) Rampage and cage data standards and processing pipeline [Online]. Available: https://www.encodeproject.org/rampage/
22. Merchant N, Lyons E, Goff S, Vaughn M, Ware D, Micklos D et al (2016) The iPlant collaborative: cyberinfrastructure for enabling data to discovery for the life sciences. PLoS Biol 14(1):e1002342
23. Stewart CA, Cockerill TM, Foster I, Hancock D, Merchant N, Skidmore E et al (2015) Jetstream: a self-provisioned, scalable science and engineering cloud environment. In: Proceedings of the 2015 XSEDE conference: scientific advancements enabled by enhanced cyberinfrastructure. XSEDE '15. ACM, New York, pp 29:1–29:8 [Online]. Available: https://doi.org/10.1145/2792745.2792774
24. Leinonen R, Sugawara H, Shumway M, International Nucleotide Sequence Database Collaboration (2011) The sequence read archive. Nucleic Acids Res 39(Database issue):D19–D21
25. Brendel VP, Raborn RT (2016) GoRAMPAGE- a workflow for promoter detection by 5'-Read mapping. https://github.com/BrendelGroup/GoRAMPAGE

26. Aronesty E (2013) Comparison of sequencing utility programs. Open Bioinform J 7(1):1–8
27. Lab H, FASTX Toolkit [Online]. Available: http://hannonlab.cshl.edu/fastx_toolkit/
28. Lassmann T (2015) TagDust2: a generic method to extract reads from sequencing data. BMC Bioinform 16(1):1
29. Li H, Handsaker B, Wysoker A, Fennell T, Ruan J, Homer N et al (2009) The sequence alignment/map format and SAMtools. Bioinformatics (Oxford, England) 25 (16):2078–2079
30. Dobin A, Gingeras TR (2016) Optimizing RNA-Seq mapping with STAR. In: Transcription factor regulatory networks. Springer, New York, pp 245–262
31. R Core Team (2017) R: a language and environment for statistical computing. R foundation for statistical computing, Vienna [Online]. Available: https://www.R-project.org
32. Lawrence M, Morgan M (2014) Scalable genomics with R and Bioconductor. Stat Sci 29 (2):214–226
33. Raborn RT, Brendel V (2017) TSRchitect: promoter identification from large-scale TSS profiling data. r Bioconductor package version 1.0.0 [Online]. Available: http://bioconductor.org/packages/release/bioc/html/TSRchitect.html
34. Yates A, Akanni W, Amode MR, Barrell D, Billis K, Carvalho-Silva D et al (2016) Ensembl (2016). Nucleic Acids Res 44:D1, D710–D716 [Online]. Available: https://doi.org/10.1093/nar/gkv1157
35. Haberle V, Forrest ARR, Hayashizaki Y, Carninci P, Lenhard B (2015) CAGEr: precise TSS data retrieval and high-resolution promoterome mining for integrative analyses. Nucleic Acids Res 43(8):gkv054–e51
36. Pagès H (2016) BSgenome: infrastructure for biostrings-based genome data packages and support for efficient SNP representation. R package version 1.42.0
37. Thorvaldsdottir H, Robinson JT, Mesirov JP (2013) Integrative genomics viewer (IGV): high-performance genomics data visualization and exploration. Brief Bioinform 14(2):178–192
38. Sayers EWE, Barrett TT, Benson DAD, Bolton EE, Bryant SHS, Canese KK et al (2012) Database resources of the national center for biotechnology information. Nucleic Acids Res 40 (Database issue):D13–D25
39. Tange O (2018) GNU parallel 2018, p 112. ISBN 978-1-387-50988-1

Chapter 10

CRM Discovery Beyond Model Insects

Majid Kazemian and Marc S. Halfon

Abstract

Although the number of sequenced insect genomes numbers in the hundreds, little is known about gene regulatory sequences in any species other than the well-studied *Drosophila melanogaster*. We provide here a detailed protocol for using SCRMshaw, a computational method for predicting *cis*-regulatory modules (CRMs, also "enhancers") in sequenced insect genomes. SCRMshaw is effective for CRM discovery throughout the range of holometabolous insects and potentially in even more diverged species, with true-positive prediction rates of 75% or better. Minimal requirements for using SCRMshaw are a genome sequence and training data in the form of known *Drosophila* CRMs; a comprehensive set of the latter can be obtained from the SCRMshaw download site. For basic applications, a user with only modest computational know-how can run SCRMshaw on a desktop computer. SCRMshaw can be run with a single, narrow set of training data to predict CRMs regulating a specific pattern of gene expression, or with multiple sets of training data covering a broad range of CRM activities to provide an initial rough regulatory annotation of a complete, newly-sequenced genome.

Key words Non-model insects, Regulatory genomics, Transcriptional gene regulation, Genome annotation, Enhancer prediction

1 Introduction

Considerable resources have been invested into insect genome sequencing over the last decade. As of May 2017 over 230 individual insect genomes had undergone some degree of genome sequencing and assembly, including orders from each of the major insect groups (Palaeoptera, Polyneoptera, Paraneoptera, Holometabola), in addition to arthropods from the Crustacea, Chelicerata, and Myriapoda. There is particular depth in the Holometabola, especially in the Hymenoptera, Diptera, and Lepidoptera (*see* http://i5k.github.io/arthropod_genomes_at_ncbi). Community efforts such as the i5k consortium seek to increase the number of sequenced arthropod genomes to as many as 5000 [1]. However, while sequence has become readily available, annotation has lagged. First-pass annotation of the coding genome is typically conducted with an automated pipeline such as MAKER [2], followed by

Susan J. Brown and Michael E. Pfrender (eds.), *Insect Genomics: Methods and Protocols*, Methods in Molecular Biology, vol. 1858, https://doi.org/10.1007/978-1-4939-8775-7_10,

manual curation to refine and/or supplement computational predictions. Subsequent improved annotation can be a labor- and resource-intensive endeavor requiring additional high-quality data from paired-end RNA-seq, directed PCR, CAGE-seq, and the like (for review of genome annotation, *see* [3, 4]). Nonprotein-coding transcription units are often poorly represented in the initial annotation, and in most cases, there is no concerted effort at all to annotate regulatory sequences. Thus a major functional genomic component—the regulatory genome—does not factor into early rounds of insect genome annotation, and with the exception of the intensively studied model insect *Drosophila melanogaster*, knowledge of insect gene regulatory sequences remains limited (for review of insect regulatory genomics, *see* [5]).

Regulation of gene transcription is a fundamental process, and detailed characterization of gene regulatory sequences represents an important step toward understanding the basic biology of any organism. In insects, as in most of the Metazoa, a significant amount of gene regulation is mediated through distal regulatory sequences that can range from immediately upstream of the promoter to hundreds of kilobases away from the transcription start site. We refer to such sequences collectively as *cis*-regulatory modules (CRMs), a classification that includes, although is not strictly limited to, what are commonly referred to as "enhancers" (for recent reviews *see* [6–11]). CRMs tend to be organized in a modular fashion, with each controlling a discrete subset of a gene's overall expression pattern. They are typically a few hundred basepairs in length and can be located 5′ to, 3′ to, within introns of, or in some cases even within exons of, their target genes. Although characterizing CRMs has historically been a laborious and low-throughput activity, the advent of genome-wide approaches coupled with next-generation sequencing technologies has led to a renaissance of sorts in methods for CRM discovery (for review *see* [8, 12–14]). However, current methods are often problematic for insects, where issues of small organism size, lack of extensive tissue-specific cell lines, and lack of mature transgenic methods for most species make these methods technically challenging and often prohibitively expensive. While the well-established community invested in *Drosophila* research has led to initiatives such as the NIH-funded modENCODE project [15] and the production of extensive experimental and computational resources, a similar infusion of funding and effort seems unlikely for non-model insects.

We present here a detailed protocol for using SCRMshaw, a computational method we developed, that leverages the wealth of available data on *Drosophila* CRMs [16] to predict regulatory sequences in both *Drosophila* and in non-model, non-drosophilid, insect species [17, 18]. SCRMshaw (for "Supervised *Cis*-Regulatory Module prediction") is effective for CRM discovery in

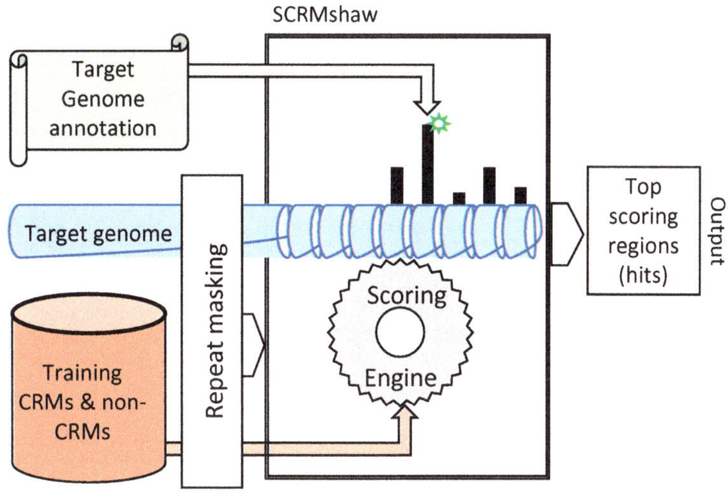

Fig. 1 Overview of the SCRMshaw *cis*-regulatory module discovery pipeline

species at least as evolutionarily distant from *Drosophila* as the 345 My diverged Hymenoptera [19]. SCRMshaw is a machine-learning based method that takes as input a sequenced insect genome along with a training set of *Drosophila* CRMs, defined by a common functional characterization (e.g., midgut expression), and outputs a set of functionally related predicted CRMs [17, 18] (Fig. 1). It does this by building statistical models that compare the short subsequence (*k*-mer) count distribution in the training CRMs to a set of randomly selected, noncoding sequences and then scoring the target genome using these models. Essentially, SCRMshaw relies on the idea that CRMs with similar function will have similar sequence characteristics, not visible by eye but identifiable using machine-learning methods, and that these regulatory mechanisms will be conserved across insect evolution. Although not universally true, we have found that these assumptions are robust enough to allow effective CRM discovery without requiring knowledge of transcription factor binding sites or of the expression patterns of the genes being regulated, between species with a virtually complete lack of observable sequence alignment at the noncoding level. Both empirical validation and comparison of predictions to regulatory regions identified by open-chromatin profiling methods suggests true-positive prediction rates averaging between 75% and 85% ([19] and MSH and Y. Tomoyasu, unpublished results). SCRMshaw can be used with all available training datasets to provide a first-pass regulatory annotation for newly sequenced insect genomes, or can be used in a targeted fashion to predict CRMs driving specific gene expression patterns of interest (e.g., [20]).

2 Materials

2.1 Genomes and Annotation Files

1. Genome sequence: Table 1 lists suggested websites for availability of insect genomes, although genomes can be downloaded from any available source (*see* **Note 1**). Minimally required is a genome FASTA file. Depending on the state of genome assembly, this may be listed simply as "genome" or as

Table 1
Resources for insect genomes and genome annotation (nonexclusive)

Resource	URL	Represented species
BIPAA: Bioinformatics platform for agroecosystem arthropods	http://bipaa.genouest.org/is/	
-aphidbase		Aphids
-lepidoBD		Lepidoptera
-ParWaspDB		Parasitoid wasps
Diamondback Moth genomic database	http://dbm.dna.affrc.go.jp/px/	*Plutella xylostella*
EnsemblMetazoa	http://metazoa.ensembl.org/index.html	Broad collection including dipterans, hymenopterans, lepidopterans, coleopterans, hemipterans, pthirapterans
FlyBase	http://flybase.org	*Drosophila melanogaster* and other drosophilids
Fourmidable	http://www.antgenomes.org	Ants
Hymenoptera Genome Database	http://hymenopteragenome.org	Bees, wasps, ants
i5k Workspace@NAL	https://i5k.nal.usda.gov	Broad collection spanning many arthropod orders
InsectBase	http://www.insect-genome.com	Broad collection spanning many arthropod orders
KAIKObase	http://sgp.dna.affrc.go.jp/KAIKObase/	*Bombyx mori*
lepbase	http://lepbase.org	Lepidoptera
SilkDB	http://silkworm.genomics.org.cn	*Bombyx mori*
SpodoBase	http://bioweb.ensam.inra.fr/spodobase/	*Spodoptera frugiperda*
Spotted Wing Flybase	http://spottedwingflybase.org	*Drosophila suzukii*
UCSC Genome Database	https://genome.ucsc.edu/	Various Diptera; many non-insects
VectorBase	https://www.vectorbase.org/	Mosquitoes and other vector species

"chromosomes," "scaffolds," or "contigs." We recommend using the most complete assembly available (*see* **Note 2**).

2. Genome Annotation: Genome annotation, minimally consisting of a set of genes and more preferably a set of genes and exons, is recommended but optional. Annotation is typically found in GFF3 format [21]. For evaluating results, it is often useful to know the *Drosophila* orthologs of the genes in your genome (*see* Subheading 3.6.4).
3. *Drosophila* genome and annotation: For creating customized training datasets (*see* Subheading 3.2), the *Drosophila melanogaster* genome and its annotation is required, along with (optionally) the genomes for related drosophilids. These can most easily be obtained from FlyBase [22] (ftp://ftp.flybase.net/genomes/). Genomes can be found in the respective "fasta/" directory for each species, and annotation in the "gff/" directory.

2.2 SCRMshaw Software

1. Operating system and software dependencies: SCRMshaw software is tested under Linux-based operating systems such as Ubuntu and CentOS. Other operating systems including Mac and Windows OS will require additional steps for installation. The core scoring methods are written in the C++ programming language and the utility scripts are implemented in Perl. The BioPerl modules "DB" and "SeqIO" are required for reading the input FASTA files. Preinstallation requirements are Perl (v5 or higher), BioPerl, and the gcc compiler (v4.8 or higher).
2. SCRMshaw and related programs: The SCRMshaw software and associated utility programs can be downloaded from http://veda.cs.uiuc.edu/SCRMshaw/ (Sinha lab) or http://halfonlab.ccr.buffalo.edu/scrmshaw.html (Halfon lab).

2.3 Training Data

SCRMshaw requires "training data" in the form of known *Drosophila* CRMs with common function (*see* **Note 3**). Precompiled sets of training data can be downloaded from the SCRMshaw website at http://veda.cs.uiuc.edu/SCRMshaw/ (Sinha lab) or http://halfonlab.ccr.buffalo.edu/scrmshaw.html (Halfon lab). These sets include, for each constituent CRM from *D. melanogaster*, orthologous sequences from 10 related *Drosophila* species; inclusion of these sequences improves SCRMshaw's predictive power [18]. Custom training sets can be developed by downloading selected sets of CRMs from the REDfly database (http://redfly.ccr.buffalo.edu; [16]) and following the steps in Subheading 3.2 (*see* Subheading 3.2 below).

2.4 Computational Requirements

1. User expertise: Although extensive computational know-how is not required for using SCRMshaw, a basic familiarity with the Linux command line is recommended. This includes being able to launch the terminal shell, navigate between directories, and

upload/download/view files. There are many online tutorials for learning basic Linux commands (e.g., https://diyhacking.com/linux-commands-for-beginners/). Programming skills are not necessary, but the source codes (in C++ and Perl) are provided for advanced users who may want to adapt the code based on their project.

2. Time and memory considerations: The computational time required for the scoring, and the required storage space, scales with the genome size. A coarse computational time estimate required per training set per million basepairs of the target genome for all three scoring methods is about 1 min on a 2.4 GHz processor. The storage space requirement for the same calculation (with the default window size = 500 bp, shift size 250 bp) is about 75 MB. Thus for a 350 MB genome, for example, computational time is ~6 h with a required 26 GB of storage. Decreasing the shift size will provide higher resolution scoring profiles; however, it will require more storage space. After the scoring, most of the storage space can be released (*see* **Note 4**). When having multiple training datasets, the computational time can be distributed among multiple processors by running an instance of SCRMshaw for each training dataset (*see* **Note 5**).

2.5 Additional Software

1. *Tandem Repeat Finder*

   Tandem Repeat Finder [23] is used for masking tandem repeat sequences and is strongly recommended for both the genome and training sequences (*see* Subheading 3.3). The correct version of Tandem Repeat Finder for your system can be downloaded from https://tandem.bu.edu/trf/trf.download.html.

2. *BEDTools*

   BEDTools [24] is a convenient and easy-to-use software suite for working with genome data. While not required for using SCRMshaw, it is highly recommended for use in associated protocols such as working with SCRMshaw output data (*see* Subheading 3.6) or creating customized training data (*see* Subheading 3.2). BEDTools can be obtained from https://github.com/arq5x/bedtools2/releases and is easily installed on a UNIX/Linux or Mac OS X computer. Documentation can be found at http://bedtools.readthedocs.io/en/latest/.

3 Methods

3.1 Installing SCRMshaw

1. Download the SCRMshaw software (*see* Subheading 2.2 above).
2. Decompress the tar archive. This will create a single "SCRMshaw" directory with the following subdirectories:

```
bin/
src/
code/
include/
example/
```

3. Navigate to the SCRMshaw directory and enter the "make" command.

Detailed information on how to install and run the software is provided in the "README.txt" file located within the software package.

3.2 Constructing Custom Training Sets

For ease of use, we recommend using precompiled training sets available on the SCRMshaw download pages (*see* Subheading 2.3). We will continue to provide updated and improved training data at these sites. However, customized training sets can also be constructed (*see* **Note 6**).

3.2.1 Obtaining CRMs from REDfly

1. Go to the REDfly database (http://redfly.ccr.buffalo.edu).
2. Use the menu bar to navigate to the "search" page and click on "Advanced Search."
3. Under "Advanced Search>Data Type" check the "CRM" button.
4. If desired, set a maximum CRM length using the "Max Size" box. We typically restrict our training CRMs to ≤2000 bp.
5. Click in the "Ontology/Expression Term" box and begin typing the name of the expression pattern to which you want to restrict the activity of the returned CRMs. An auto-complete drop-down list will enable you to select the desired term. Alternatively, you can enter a *Drosophila* Anatomy Ontology ID number (FBbt:xxxxxxxx). See the REDfly online "Help" for additional information on expression pattern searching (also *see* **Note 7**).
6. Click "Search."
7. A list of CRMs meeting the search criteria will appear in the "Search Results" area. Click the box on the left of the header bar to select all CRMs, then click "Download Selected" at the bottom of the page.
8. In the dialogue box, select "File Type: FASTA," "Coord Version: R6," and "FASTA Options: Sequence Only." Click download.
9. The downloaded file contains a multi-FASTA file with the *Drosophila melanogaster* members of the training set. This can be used as is or checked for overlapping sequences and refined (*see* **Note 8**) before moving on to obtain orthologous

sequences (*see* Subheading 3.2.2) and generating the negative training data (*see* Subheading 3.2.3). To check for overlapping sequences and/or to obtain orthologous CRMs, it is convenient to also download the same selected set of CRMs in BED format. This is accomplished by following the same steps as in **step 8**, above, substituting "File Type: BED" and "BED File Type: BED simple."

3.2.2 Obtaining Orthologous Regions from Other Drosophila *Species*

Many of the sequenced *Drosophila* species are evolutionarily close enough to *D. melanogaster* (within up to 40 My of divergence) that there is a high level of conservation of their noncoding genomes, and as such extensive alignments of noncoding regions can be constructed. We have shown that augmenting the training set CRMs with the orthologous sequences from additional species improves SCRMshaw performance [18]. (Presumably this is because the most important *k*-mers within the CRMs are under evolutionary constraint and therefore more conserved than background *k*-mers, boosting the weight of the most relevant *k*-mers within the training data.) Orthologous CRMs can be obtained using the *liftOver* tool at the UCSC Genome Browser [25]:

1. Obtain the FASTA and BED files for the training CRMs from **steps 8** and **9** (*see* Subheading 3.2.1 above) and rename them "crm.fa" and "crm.bed," respectively.
2. Navigate to the liftOver tool at https://genome.ucsc.edu/cgi-bin/hgLiftOver (*see* **Note 9**).
3. For "Original Genome" select *D. melanogaster*. Choose an appropriate Original Assembly, "New Genome," and New Assembly from the drop-down lists (*see* **Note 10**).
4. [Optional] To increase the stringency of mapping between genomes, change the "Minimum ratio of bases that must remap" from 0.1 (default) to 0.25.
5. Upload the "crm.bed" file obtained in **step 9** (*see* Subheading 3.2.1 above) using "Choose File," followed by "Submit."
6. Download and save the successful conversions from the link "View Conversions" as "species1_crms.bed."
7. Download the correct genome assembly for the "New genome" ("species1_genome.fa"; *see* Subheading 2.1, **item 3**).
8. Use "bedtools getfasta" from the BEDTools software suite [24] to obtain the FASTA formatted sequences for the CRMs from the orthologous species:

```
$bedtools getfasta -fi species1_genome.fa -bed species1_crms.bed -fo species1_crms.fa
```

9. Repeat **steps 2–6** for all other orthologous species of interest.
10. Combine the FASTA files to create one training CRM file, e.g.:

```
$cat species*_crms.fa crm.fa > project1/dataset1/crms.
fasta
```

3.2.3 Creating the Negative Training Sets (Non-CRMs, or Background Sequences)

Two of our scoring methods (HexMCD and IMM) maximize the likelihood ratio between the similarity of a target sequence to the training CRMs versus its similarity to a "background set" of non-CRMs, in terms of their *k*-mer profiles. The ideal negative training set should be depleted of the *k*-mers that distinguish CRMs from non-CRMs, while retaining all the other *k*-mers (*see* **Note 11**).

To obtain an appropriate set of randomly selected noncoding sequences to use as the background set, we have provided a script called "randomWithSameGC.pl" in the SCRMshaw package that extracts random regions in the genome with the same GC content as the CRMs (±1%). This script can be run from within the SCRMshaw directory as:

```
$perl code/randomWithSameGC.pl --crm crms.fasta --output neg.
fasta --size 10 --genomedir rootDIR --gene gene.gff3
```

The genome root directory ("rootDIR") is the path to the genome FASTA file (i.e., the *D. melanogaster* genome, *see* Subheading 2.1, **item 3** above). For example, if the genome is located at "project1/genome.fa" then the root directory is "project1/." The size option defines the number of background CRMs matched to each training CRM (*see* **Note 12**). The gene annotation file is used to filter out regions that overlap coding sequences.

3.3 Masking Tandem Repeats

Tandem repeats, as the name indicates, consist of a repeated pattern of one or more nucleotides that occur directly after each other. An example would be ACACACACAC, where the dinucleotide "AC" is repeated five times. The occurrence of a long tandem repeat in the training dataset or the region to be scored significantly skews the distribution of *k*-mer counts toward the repeated pattern, which would result in assignment of a "false" high score to regions with one or more occurrences of the repeated pattern. To avoid this potential issue, tandem repeats in both the training datasets and the genome to be searched are masked prior to running SCRMshaw using Tandem Repeat Finder [23]. To mask tandem repeats:

1. Navigate to "https://tandem.bu.edu/trf/trf.download.html" and download the current version of TRF for your system type (e.g., trf409.linux64).
2. Change the downloaded TRF file to an executable form:

```
$chmod +x trf409.linux64
```

3. Run TRF on your FASTA files (e.g., genome.fa, crms.fa, neg.fa) using parameters as shown:

```
$trf409.linux64 genome.fa 2 7 7 80 10 50 500 -m -h
$trf409.linux64 crms.fasta 2 7 7 80 10 50 500 -m -h
$trf409.linux64 neg.fasta 2 7 7 80 10 50 500 -m -h
```

4. Move the TRF output to your training set directory and rename:

```
$mv genome.fa.2.7.7.80.10.50.500.mask project1/genome.fa
$mv crms.fasta.2.7.7.80.10.50.500.mask project1/dataset1/crms.fasta
$mv neg.fasta.2.7.7.80.10.50.500.mask project1/dataset1/neg.fasta
```

TRF substitutes the repeated nucleotides with "N" characters that can be directly fed into SCRMshaw. If using a different tool for masking repeats, be sure that the repeated nucleotides are replaced by the same number of "N" characters.

3.4 Running SCRMshaw

SCRMshaw requires the following files and directories:

1. A genome file (e.g., "genome.fa"), a multi-FASTA file including all of the genomic sequences needing to be scored. This is typically a downloaded genome file (*see* Subheading 2.1, **item 1**) that has been masked for tandem repeats (*see* Subheading 3.3 above), and is passed to the program using the "--genome" option.
2. One or more dataset directories each containing two files, "crms.fasta" and "neg.fasta." These files are respectively the positive and negative training sequences, both in FASTA file format, as described above (*see* Subheadings 2.3 and 3.2).
   We highly recommend tandem repeat masking of genome.fa, crms.fasta, and neg.fasta (*see* Subheading 3.3 above).
3. A text file containing a list of the dataset directories (e.g., "trainingSet.lst"; *see* below **step 4**). This is specified using the "--traindirlst" option.

To run SCRMshaw:

1. Create a project directory (e.g., project1/):

```
$mkdir project1/
```

2. Under the project directory, create one or more dataset directories (here, dataset1/, dataset2/, etc.) (*see* **Note 13**):

```
$mkdir project1/dataset1/
```

3. Move each pair of positive and negative training datasets to the appropriate dataset directory:

```
$mv crms.fasta project1/dataset1/
$mv neg.fasta project1/dataset1/
```

4. Under the project directory, create a list file "trainingSet.lst" that contains the path to the datasets (e.g., project1/dataset1/), one line per dataset. You can create this file using a simple text editor or using the following "echo" commands:

```
$echo "project1/dataset1/" >> project1/trainingSet.lst
$echo "project1/dataset2/" >> project1/trainingSet.lst
$echo "project1/dataset3/" >> project1/trainingSet.lst
```

etc.

5. Move your gene annotation file to the project directory.

```
$mv gene.gff3 project1/
```

6. For simplicity, move the genome FASTA file to your project directory:

```
$mv genome.fa project1/
```

At the conclusion of these steps, your directory will resemble Fig. 2.

7. Run SCRMshaw, e.g.:

```
$perl code/scrm.pl --thitw 2000 --gff project1/gene.gff3
--imm --hexmcd --pac --genome project1/genome.fa --train-
dirlst project1/trainingSet.lst --outdir project1/results/
--step 123
```

The SCRMshaw results will be placed in the directory specified by "--outdir," which will be generated automatically if it doesn't exist (here, "project1/results/").

The SCRMshaw options include:

(a) the name of scoring method(s) to be used for prediction ("--imm" for IMM, "--hexmcd" for HexMCD, and "--pac" for PAC): required. Any individual or combination of methods can be selected to run (e.g., "--imm --pac" to run IMM and PAC). For details about the individual scoring methods *see* [17, 18].

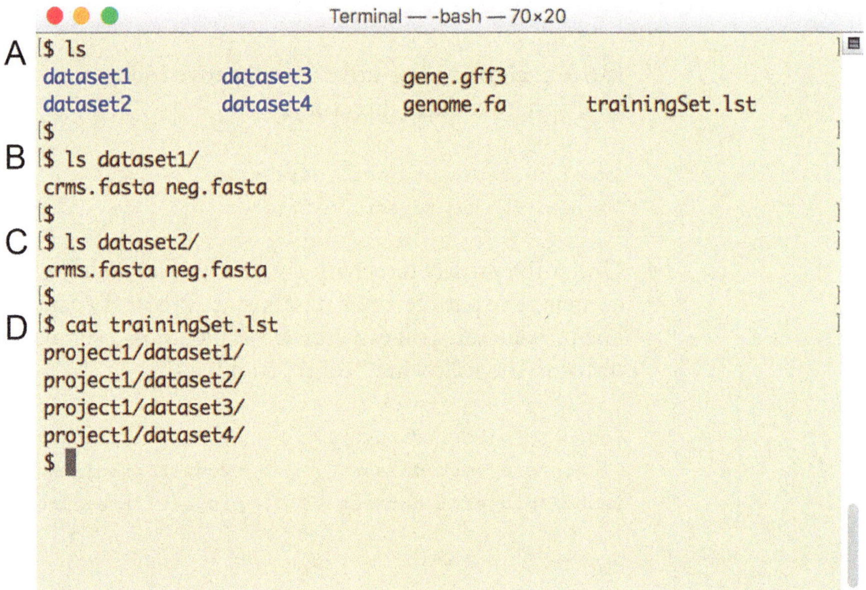

Fig. 2 Preparing to run SCRMshaw. The screenshot illustrates the directory setup and required user-added files for using SCRMshaw. (**a**) Directory structure showing the genome file, annotation file, training set list file, and training set subdirectories. (**b**, **c**) Each training set subdirectory contains a positive (crms.fasta) and negative (neg.fasta) training data file in the form of a pair of multi-FASTA sequence files. (**d**) The "trainingSet.lst" file is a text file containing a list of paths to the training set subdirectories

(b) the gene annotation file of the genomic regions to be scored in GFF3 format (e.g., "--gff gene.gff3"): optional but strongly recommended. Alternatively, separate "gene" and "exon" files can be used (*see* **Note 14**). It is also possible to use SCRMshaw without providing any annotation, but this will not allow for exclusion of coding sequences or calculation of "local rank" (*see* below).

(c) the number of top scoring regions to be reported as CRM predictions (--thitw N): optional, default = 2000 (*see* **Note 15**).

(d) which steps of the SCRMshaw pipeline to run (--step 123): optional, default = 123. **Step 1** processes the genome and genome annotation, **step 2** processes the training data, and **step 3** scores the genome (*see* **Note 5**).

Detailed information on how to run SCRMshaw from the command line and additional available options can be obtained by the following command: "perl code/scrm.pl" or from the README document accompanying the SCRMshaw distribution. An example benchmark data set along with instructions is provided in the example directory within the software package.

3.5 The SCRMshaw Output

SCRMshaw provides several pieces of information about each predicted CRM, or "hit." These include:

(a) *Chromosome_ID:start_coordinate.* This is the name of the chromosome/scaffold/contig and the nucleotide coordinate of the start of the scored window. The end coordinate can be determined by simply adding the length of the window to the start coordinate.

(b) *Score.* The score for the window (*see* **Note 16**).

(c) *Gene_ID:Gene_symbol:distance:location.* The *Gene_ID* and *Gene_symbol* are drawn from the genome annotation (these will sometimes be identical, depending on how the annotation was constructed). *distance* provides the distance in base pairs between the start of the window and the annotated gene, while *location* takes the value "upstream," "downstream," or "inside" depending on whether the window is 5′, 3′, or within an intron of the gene (for windows within introns, *distance* is given as 0).

(d) *local_rank:* The local rank refers to the rank score of the window considering only a 50 kb range to either side of the annotated gene [19]. (This distance can be set to a different value at run time using the "--distance" option.) Local ranks are only provided when an annotation file is used (*see* **Note 16**).

The SCRMshaw output is placed in the directory specified by the "--outdir" option and has the following subdirectories (Fig. 3a):

(a) *hits/.* This is the main subdirectory of interest and contains the top hits (i.e., predicted CRMs) for each of the methods for each training dataset. A separate subdirectory is created for each method (*hexmcd, imm, pac*). Each contains a subdirectory for each training set, with three "hits" files (Fig. 3b):

dataset1.hits. This file contains the top-scored windows in rank order in FASTA format:

```
>chromosome_ID:start_coordinate score gene_ID:gen-
e_symbol:distance: location
ATGCCCAGAGAATGGGCAACAAGTAGCGGCGAATTAGCAATCCTAT-
CATGCTTTTATGGCCGGCCAACTCTTGCC
```

dataset1.hits.ranked. This file contains the top-scored windows in rank order in FASTA format, but with local ranks also included:

```
>chromosome_ID: start_coordinate score gene_ID:
gene_symbol:distance: location:local_rank AATGCCCA
GAGAATGGGCAACAAGTAGCGGCGAATTAGCAATCCTATCATGCTTT
TATGGCCGGCCAACTCTTGCC
```

A

```
___ project1/results/
                    |
                    |___ fasta/
                    |         |
                    |         |___ chr/
                    |         |
                    |         |___ kmers/
                    |         |
                    |         |___ windows/
                    |
                    |___ gff/
                    |
                    |___ hits/
                    |        |
                    |        |___ hexmcd/
                    |        |
                    |        |___ imm/
                    |        |
                    |        |___ pac/
                    |
                    |___ scores/
                    |          |
                    |          |___ hexmcd/
                    |          |
                    |          |___ imm/
                    |          |
                    |          |___ pac/
                    |
                    |___ training/
```

B

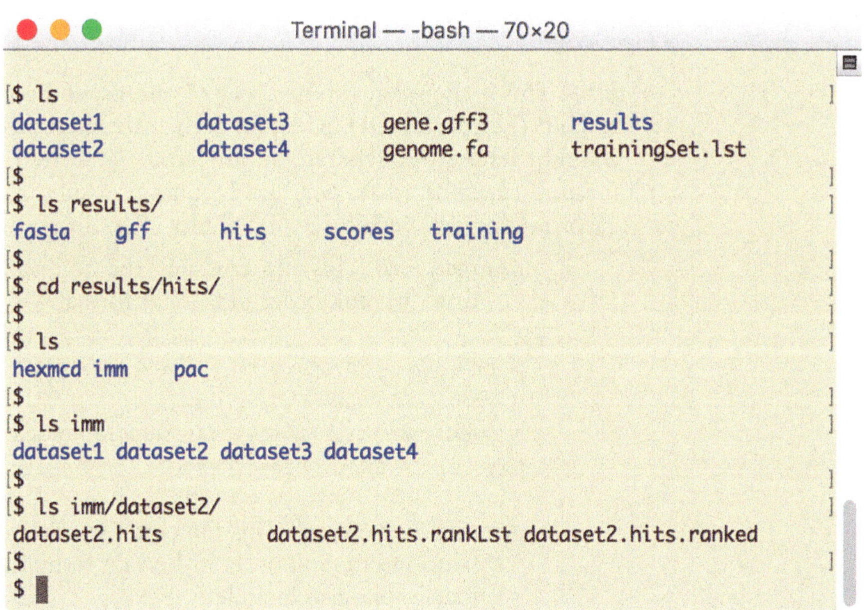

Fig. 3 SCRMshaw output. (**a**) Outline of the subdirectory structure established by SCRMshaw in the assigned "results" directory. (**b**) Screenshot showing the various results subdirectories and their component files

dataset1.hits.rankLst. This file contains the top-scored windows without sequence in tabular format:

Gene_ID, chromosome_ID: start_coordinate, score, local_rank

If no GFF file or gene/exon annotation was provided, only the "dataset1.hits" file will be generated.

(b) *fasta/*. This directory has three subdirectories containing intermediate files used by SCRMshaw during the run. These are:

chr/: This directory stores the target sequence FASTA files, one sequence per file. Exon regions will be masked if GFF or exon files were provided.

kmers/: This directory stores *k*-mer frequency files for each target sequence.

windows/: This directory contains the target sequence with each chromosome/scaffold/contig stored as a multi-FASTA file broken into window-length fragments.

(c) *gff/*. This directory contains gene and exon locations extracted from the provided GFF annotation file.

(d) *scores/*. This directory contains the raw scores of a sliding window across the target genome, with one score file per chromosome/scaffold/contig. Each score file contains two columns, the location of the window and its corresponding score. Scores are contained in a separate subdirectory for each method (*hexmcd, imm, pac*).

(e) *training/*. This directory contains intermediate files with the word statistics of the training sets that are used for building the statistical models and scoring the target genome.

Once the results have been obtained, storage space can be freed up by keeping the "hits" files but discarding the various intermediate files (*see* **Note 4**).

3.6 Working with the SCRMshaw Output

In this section, we provide some protocols and tips for making productive use of the SCRMshaw output.

3.6.1 Getting Top Hits in BED Form

For many applications, it will be useful to generate a list of the top SCRMshaw results of interest as a BED-compatible file. We have developed a simple Perl utility script, "Generate_top_N_SCRMhits," to facilitate this task. The script takes a user-defined number "*N*" of desired results and outputs the top-ranked *N* predicted CRMs for each training set and each method (*hexmcd, imm, pac*) found in the /hits directory from a given set of SCRMshaw runs. Output is in the form of a 16-column tab-delimited file as follows:

1. chromosome
2. start
3. end
4. SCRMshaw score
5. flanking gene
6. flanking gene (due to a quirk in the way SCRMshaw results are reported, this will often be a repeat of field 5)
7. distance of hit from flanking gene (basepairs)
8. location of hit relative to flanking gene
9. local rank
10. next closest flanking gene
11. next closest flanking gene (due to a quirk in the way SCRMshaw results are reported, this will often be a repeat of field 10)
12. distance of hit from flanking gene (basepairs)
13. location of hit relative to flanking gene
14. local rank
15. training set
16. method (*hexmcd*, *imm*, *pac*)

Run the script from the command line with the following options:

-d directory, path to directory containing SCRMshaw results (i.e., directory name provided as "--outdir" when running SCRMshaw): *required*

-o outfile, name of file for results of script: *required*

-n number of hits desired: *defaults to 250 (for each training set and each method)*

-h help, displays usage information: *overrides other options*

-v version, displays program version: *overrides other options*

3.6.2 Sorting and Merging Hits

It may be useful, once having generated the list of top predicted CRMs, to merge together overlapping regions and/or duplicate predictions (i.e., made from more than one method). This is most easily accomplished using BEDTools [24] as follows:

```
$bedtools sort -i Top_N_Hits_file | bedtools merge -c 4,5,10,15,16 -o max,distinct,distinct,distinct,distinct
```

or

```
$sort -k1,1 -k2,2n Top_N_Hits_file | bedtools merge -c 4,5,10,15,16 -o max,distinct,distinct,distinct,distinct
```

The resulting file will be an 8-column tab-delimited file with columns:

1. chromosome
2. start
3. end
4. SCRMshaw score (if predicted CRMs were merged, the highest score from the merged set)
5. flanking gene
6. next closest flanking gene
7. training set (if predicted CRMs were merged from multiple training sets, a comma-separated list of the training sets)
8. method (if predicted CRMs were merged from multiple methods, a comma-separated list of the methods)

3.6.3 Comparing SCRMshaw Predictions to Other Genomic Data Sets

It may often be of interest to compare SCRMshaw results with sets of known CRMs, other sets of predicted CRMs, or other genomic data such as histone modifications, transcription factor binding, or open-chromatin regions. This can be done quite simply using the merged top-hits file generated from the above protocol (*see* Subheadings 3.6.1 and 3.6.2) and the BEDTools program [24]. We recommend setting a minimum overlap of 10% (−f 0.10) for the predicted CRMs. An unmerged predicted CRM would thus need to overlap another genomic feature by 50 bp, while a longer merged CRM would require 75 bp of overlap (in our experience it is rare for merged features to exceed 750 bp). A typical command for comparing SCRMshaw results to another set of genomic features would be:

```
$bedtools intersect -loj -f 0.10 -a merged_Top_N_hits_file -b
other_genomic_features_file
```

3.6.4 Enrichment Test

One difficulty when using SCRMshaw to predict CRMs for a newly sequenced, non-model insect genome, where there may be few or no known CRMs to validate against, is determining whether SCRMshaw worked effectively, i.e., with a low false-positive rate. One way to attempt a rough assessment of this is to find out whether the predictions in the target species fall near the orthologs of the training set genes, or other genes whose expression is consistent with that of the training set genes (e.g., for a "salivary gland" training set, a set of genes all expressed in the salivary glands). Although orthologous genes do not always maintain a similar expression pattern, such an analysis can at least provide an indirect confirmation of the results. This sort of "enrichment test" analysis is described in [19], and the necessary software can be obtained from the SCRMshaw download page (click on "pipeline for enrichment test").

Orthologous gene mappings can be obtained from a variety of sources including OrthoDB (http://www.orthodb.org; [26]); InParanoid (http://inparanoid.sbc.su.se/; [27]); and EggNOG (http://eggnogdb.embl.de/; [28]). Precompiled lists of orthologous genes between *Drosophila* and many other species can be downloaded from InParanoid (http://inparanoid.sbc.su.se/download/current/Orthologs_other_formats/D.melanogaster/) or accessed via FlyBase.

4 Notes

1. SCRMshaw can be run with any sequenced insect genome; however, to date we have only used genomes from within the Holometabola. Although empirical evaluations have not been extensive enough to make a definitive determination, initial results suggest that true-positive CRM prediction rates are similar for holometabolous insects regardless of whether they are evolutionarily closer to (e.g., mosquitoes) or further from (e.g., bees, wasps) *Drosophila* [19, 20].
2. That is, we recommend favoring chromosome > scaffold > contig. However, the effect of genome assembly quality on SCRMshaw performance has not yet been determined, and we cannot, at this time, suggest a minimal acceptable scaffold N50 or gap percentage (we are currently developing simulations to address this question). Because by default SCRMshaw scores the genome using 500 bp windows, we do not expect the extent of genome assembly to have a major role in determining the effectiveness of CRM prediction.
3. The "common function" of a set of CRMs can be defined in various ways, and broadly or narrowly. Ideal characteristics are an ongoing area of investigation. As a general rule, the most successful training datasets consist of CRMs with similar (although not necessarily identical) patterns based on well-defined anatomical classifications such as "somatic muscle," "wing disc," and "ventral nerve cord." We expect that the more broadly defined the class, the wider the functional range of the resulting CRM predictions will be. This allows the user to search for a more-or-less specific set of results. We have found that even very broadly constituted datasets give a higher percentage of true-positive results than training data composed of randomly selected noncoding sequences, suggesting that as a whole CRMs are enriched for informative subsequences [17, 29].

4. SCRMshaw creates a large number of files when processing the genome and training data, which contain breakdowns of individual chromosome/scaffold FASTA sequences (/fasta/chr), details on *k*-mer frequencies (/fasta/kmers), and FASTA sequences of each 500 bp window used in scoring (/fasta/windows). For a large genome, these files can require significant storage space, but they are not required for routine applications once the predicted CRM results have been obtained. We provide a simple utility shell script on the SCRMshaw website, "cleanup_fastadirectory.sh," that will delete the contents of the /fasta directory to free up storage space.
5. For a given genome, we recommend first running SCRMshaw with a limited number of training datasets and the "--step" option set to "--step 123" (*see* Subheading 3.4, **step** 7). This will process the genome and annotation, and ensure that all of the proper subdirectories are created. Parallel instances of SCRMshaw can then be run with additional training sets and "--step 23." To run an additional scoring method for a previously run training set (e.g., if only ran "--imm" initially and now want to add "--hexmcd"), only "--step 3" is necessary.
6. Training set quality has a significant impact on CRM prediction. Presently, we have no method in place for evaluating training set sensitivity (how many known CRMs are recovered), while the specificity of existing training sets (how many of the predicted CRMs are likely to have the expected regulatory activity) can be inferred from the cross-validation and empirical testing data reported in [17, 18]. New training sets can be evaluated using the "enrichment test" protocol (*see* Subheading 3.6.4). However, it can be difficult to distinguish by this test if poor performance is due to the training set or to the insect genome being studied. We are developing a comprehensive suite of training set evaluation tests that will allow for assessment of both sensitivity and specificity of new and existing training sets. These tools will be made available on the SCRMshaw website when completed.
7. REDfly presently allows only a single expression term to be searched. If multiple search terms are desired, search each term individually using the listed procedure, then combine the download results into a single file, manually removing any duplicate CRMs.
8. It is desirable to remove any sequence overlaps, as including the same sequence twice will artificially weight the *k*-mers belonging to that sequence. This can be achieved by first sorting the BED file of REDfly CRMs (e.g., using BEDTools sort) and then running the utility Python script "SelectSmallestFeature.py," available for download at the SCRMshaw and REDfly websites.

9. LiftOver can also be downloaded and installed locally, following instructions at https://genome-store.ucsc.edu. All options can be used as default with the exception of "minMatch," which we recommend changing to 0.25.
10. The current version of the *D. melanogaster* genome is release 6, listed at the UCSC Genome Browser site as August 2014/dm6. However, few other genomes are mappable using liftOver for this release; most of the relevant other drosophilids are only available through the April 2006/dm3 genome version (release 5). REDfly CRMs through at least REDfly v5.2.1 can be downloaded as release 5/dm3 coordinates and in that form used directly to get orthologous regions via liftOver. Alternatively, a list of REDfly CRM coordinates can be converted from release 6/dm6 to release 5/dm3 using either liftOver or the conversion tool at FlyBase (http://flybase.org/static_pages/downloads/COORD_625.html). While these conversions can be used to obtain FASTA sequences for compiling training data, we recommend using the most up-to-date genome release and annotation when running SCRMshaw.

    We used the following ten genomes in our previous work [18]: *D. ananassae, D. erecta, D. grimshawi, D. mojavensis, D. persimilis, D. pseudoobscura, D. sechellia, D. simulans, D. virilis,* and *D. yakuba*.
11. Choosing the "right" negative training set is important for the scoring. One strategy to obtain negative sequences would be to randomly shuffle the nucleotides within CRMs. The negative sequences from this strategy would have the same GC content as the CRM set, but would likely be depleted in the higher order *k*-mer features shared by the CRMs. However, due to random shuffling they might also inadvertently lack *k*-mers that are common to the non-CRMs. We prefer to select random noncoding regions in the source (*D. melanogaster*) genome having the same GC content as the CRMs ($\pm 1\%$). This alleviates the inadvertent effect of the above-mentioned strategy. On the other hand, it is likely that these sequences will contain other, unknown CRMs. Nevertheless, our previous results suggest that this method provides effective negative training data. The script "randomWithSameGC.pl," included in the SCRMshaw package, can select appropriate sequences.

    Note, however, that we have not fully explored the effect of various negative training sets on SCRMshaw performance. Other possibilities include obtaining the negative training sequences from the target genome (i.e., the genome that will be scored), or using the known *Drosophila* CRMs that do not belong to the positive training set. A version of this latter is used in the "IMMBoost" method recently reported by Yang and Sinha [30]. IMMBoost appear to increase specificity of

CRM discovery, but effects on sensitivity and use in a cross-species (i.e., scoring non-*Drosophila* genomes) have not yet been evaluated.

12. We typically use a tenfold excess of negative training sequences (i.e., --size 10). Although we have not explored the effects of varying this parameter, our assumption is that the larger the negative set, the more accurate the statistics of the *k*-mer distribution.
13. If using SCRMshaw in a multi-user setting, other directory structures might be more practical. For example, all of the training datasets can be placed in a single "datasets" directory outside of the "project" directory where they will be accessible for multiple projects. Paths in the "trainingSet.lst" file (**step 4**) should then be adjusted accordingly (e.g., "../datasets/dataset1").
14. Individual "genes" and "exons" files can be used in lieu of an annotation GFF3 file; one or both files can be provided. These should be tab-delimited lists of gene or exon coordinates, respectively, in the form chromosome (or scaffold), start coordinate, end coordinate, strand (+ or −), gene ID/exon ID (e.g., gene_name:1). In the absence of an annotation GFF3 file, running SCRMshaw without a "genes" file will preclude calculation of local rank and provision of results indicating closest flanking genes. Omitting an "exons" file will prevent masking of exons and allow coding regions to be scored as potential CRMs.
15. The number of predictions to be reported can be adjusted depending on expectations of how many CRMs might be recovered, for instance based on how functionally broad or specific the training CRMs are. At the moment, we do not have a systematic way of learning this parameter from the input training sets. We recommend using a generous value, and then if desired working with a smaller subset of results obtained using the "Generate_top_N_SCRMhits" script (*see* Subheading 3.6.1).
16. Scores are used in determining the "global rank" of each hit [19] using the rule of thumb that higher scores are stronger predictions than lower scores. However, we do not at present have a clear guideline for "good" versus "poor" absolute scores. In [19], an msIMM score threshold of four provided good discrimination between pattern-specific (expression matching the training data) and nonspecific predicted CRMs, but our overall experience suggests that this number does not generalize. Negative-scoring windows from IMM and HexMCD should be avoided, as the negative score indicates

they are more similar to the background than the training CRMs.

"Local rank" refers to the rank of a window score with respect only to the 50 kb region to either side of an annotated gene (more strictly, with respect to the region specified by the "--distance" option) [19]. In general, we would expect a true CRM that activates a nearby gene in a manner consistent with the training data to fall within one of the highest-scoring windows in the locus. (As genes may have multiple similar CRMs [31], a true-positive prediction does not necessarily need to be the top-ranked local window.)

Acknowledgments

We thank Kushal Suryamohan for comments on the manuscript. This work was supported by USDA grant 2012-67013-19361 (MSH) and NIH grant 5K22HL125593-02 (MK).

References

1. i5k Consortium (2013) The i5K Initiative: advancing arthropod genomics for knowledge, human health, agriculture, and the environment. J Hered 104:595–600
2. Holt C, Yandell M (2011) MAKER2: an annotation pipeline and genome-database management tool for second-generation genome projects. BMC Bioinformatics 12:491
3. Ekblom R, Wolf JB (2014) A field guide to whole-genome sequencing, assembly and annotation. Evol Appl 7:1026–1042
4. Yandell M, Ence D (2012) A beginner's guide to eukaryotic genome annotation. Nat Rev Genet 13:329–342
5. Suryamohan K, Halfon M (2015) Insect regulatory genomics. In: Raman C et al (eds) Short views on insect genomics and proteomics. Springer International Publishing, pp 119–155
6. Cho, K.W. (2012) Enhancers. Wiley interdisciplinary reviews developmental biology, vol. 1, pp 469–478
7. Long HK et al (2016) Ever-changing landscapes: transcriptional enhancers in development and evolution. Cell 167:1170–1187
8. Shlyueva D et al (2014) Transcriptional enhancers: from properties to genome-wide predictions. Nat Rev Genet 15:272–286
9. Smith E, Shilatifard A (2014) Enhancer biology and enhanceropathies. Nat Struct Mol Biol 21:210–219
10. Vernimmen D, Bickmore WA (2015) The hierarchy of transcriptional activation: from enhancer to promoter. Trends Genet 31:696–708
11. Buffry AD et al (2016) The functionality and evolution of eukaryotic transcriptional enhancers. Adv Genet 96:143–206
12. Suryamohan K, Halfon MS (2015) Identifying transcriptional cis-regulatory modules in animal genomes. Wiley Interdiscip Rev Dev Biol 4:59–84
13. Li Y et al (2015) The identification of *cis*-regulatory elements: a review from a machine learning perspective. Biosystems 138:6–17
14. Murakawa Y et al (2016) Enhanced identification of transcriptional enhancers provides mechanistic insights into diseases. Trends Genet 32:76–88
15. modENCODE Consortium et al (2010) Identification of functional elements and regulatory circuits by Drosophila modENCODE. Science 330:1787–1797
16. Gallo SM et al (2011) REDfly v3.0: toward a comprehensive database of transcriptional regulatory elements in Drosophila. Nucleic Acids Res 39:D118–D123
17. Kantorovitz MR et al (2009) Motif-blind, genome-wide discovery of cis-regulatory modules in Drosophila and mouse. Dev Cell 17:568–579

18. Kazemian M et al (2011) Improved accuracy of supervised CRM discovery with interpolated Markov models and cross-species comparison. Nucleic Acids Res 39:9463–9472
19. Kazemian M et al (2014) Evidence for deep regulatory similarities in early developmental programs across highly diverged insects. Genome Biol Evol 6:2301–2320
20. Suryamohan K et al (2016) Redeployment of a conserved gene regulatory network during Aedes aegypti development. Dev Biol 416:402–413
21. Stein, L. (2013) Generic Feature Format Version 3 (GFF3). https://github.com/The-Sequence-Ontology/Specifications/blob/master/gff3.md
22. Gramates LS et al (2017) FlyBase at 25: looking to the future. Nucleic Acids Res 45: D663–D671
23. Benson G (1999) Tandem repeats finder: a program to analyze DNA sequences. Nucleic Acids Res 27:573–580
24. Quinlan AR, Hall IM (2010) BEDTools: a flexible suite of utilities for comparing genomic features. Bioinformatics 26:841–842
25. Kent WJ et al (2002) The human genome browser at UCSC. Genome Res 12:996–1006
26. Zdobnov EM et al (2017) OrthoDB v9.1: cataloging evolutionary and functional annotations for animal, fungal, plant, archaeal, bacterial and viral orthologs. Nucleic Acids Res 45:D744–D749
27. Sonnhammer EL, Ostlund G (2015) InParanoid 8: orthology analysis between 273 proteomes, mostly eukaryotic. Nucleic Acids Res 43:D234–D239
28. Huerta-Cepas J et al (2016) eggNOG 4.5: a hierarchical orthology framework with improved functional annotations for eukaryotic, prokaryotic and viral sequences. Nucleic Acids Res 44:D286–D293
29. Suryamohan, K. (2016) PhD Thesis: Regulatory networks in development: understanding the role of *cis*-regulatory modules in Gene Regulatory Network evolution. Department of Biochemistry, University at Buffalo-State University of New York
30. Yang W, Sinha S (2017) A novel method for predicting activity of cis-regulatory modules, based on a diverse training set. Bioinformatics 33:1–7
31. Barolo S (2012) Shadow enhancers: frequently asked questions about distributed cis-regulatory information and enhancer redundancy. Bioessays 34:135–141

Chapter 11

Whole-Genome Bisulfite Sequencing for the Methylation Analysis of Insect Genomes

Fanny Gatzmann and Frank Lyko

Abstract

DNA methylation is a conserved epigenetic modification of animal genomes, but genome methylation patterns appear surprisingly diverse in insects. Whole-genome bisulfite sequencing (WGBS) represents a sensitive and robust method for the characterization of genome-wide methylation patterns at single-base resolution. Here, we describe a step-by-step protocol for the generation and analysis of WGBS datasets using standard Illumina sequencing platforms. In comparison to whole-genome sequencing, WGBS has additional caveats that require particular attention and are highlighted in this chapter.

Key words Epigenetics, DNA methylation detection, Whole-genome bisulfite sequencing, WGBS

1 Introduction

DNA methylation is a biochemical process that predominantly involves the covalent addition of a methyl group to the carbon-5 of cytosines and is catalyzed by the family of DNA methyltransferases [1]. In animals, DNA methylation is targeted mainly to CpG dinucleotides and plays a crucial role in a variety of biological processes including gene regulation, embryonic development, chromatin remodeling, and suppression of transposable element activity [2–4]. DNA methylation has also been suggested to function as an adaptive mechanism that allows the organism to respond to environmental influences without altering the genome sequence [5]. However, experimental data supporting this hypothesis remains sparse and is often discussed controversially.

DNA methylation is present in all domains of life, and the patterns of DNA methylation can show substantial diversity [6, 7]. Variation of DNA methylation patterns appears particularly pronounced in insects and has been shown to range from dense methylation in locusts [8] to sparse methylation in honeybees [9]. Moreover, some insect genomes appear to have lost DNA

Susan J. Brown and Michael E. Pfrender (eds.), *Insect Genomics: Methods and Protocols*, Methods in Molecular Biology, vol. 1858, https://doi.org/10.1007/978-1-4939-8775-7_11, © Springer Science+Business Media, LLC, part of Springer Nature 2019

methylation completely [10, 11], suggesting context-dependent evolution [12].

Bisulfite sequencing represents the gold-standard methodology to analyze methylation patterns at single-base resolution [13]. DNA is treated with sodium bisulfite, which leads to selective deamination of unmethylated cytosines, followed by a conversion to uracil. After sequencing, unmethylated cytosines will appear as thymines, while only methylated and hydroxymethylated [14] cytosines will appear as cytosines. More recently, high-throughput sequencing technologies have enabled the generation of genome-wide methylation maps at single-base resolution [15]. This method, termed whole-genome bisulfite sequencing (WGBS) continues to have a transformative influence on our understanding of DNA methylation. Due to their highly diverse methylation patterns, insects are particularly interesting for studying functional aspects of DNA methylation in an evolutionary context [16].

We describe here a protocol for the analysis of global DNA methylation patterns by whole-genome bisulfite sequencing using standard next-generation Illumina platforms. DNA isolation and library preparation are based on commercially available kits that include detailed instruction manuals. Sequencing reads are computationally processed and mapped to the reference genome of the organism. Results can be viewed and analyzed as Genome Browser tracks and by customized tools for WGBS data analysis.

2 Materials

2.1 Hardware Requirements and System Options

Handling the large raw sequencing data files is generally computing-power and memory demanding. If possible, we recommend using a high-performance computing (hpc) infrastructure, and if that is not possible, a powerful desktop computer. Most NGS software are released and/or optimized for Linux, and we give usage and installation guides below for Linux operating systems. If you are working with a Windows PC, please consult Chapter 12 [17]. Like Asselman (Chapter 12 [17]), we recommend at least 16Gb RAM and preferably 5 cores. All software introduced here is publicly available.

2.2 Software Tools and Quick Installation Guides

1. Java (java.com)—Some of the software tools depend on a suitable Java Runtime Environment (JRE). To set this up, visit java.com and click on "Free Java Download." Click "See all java downloads." Find your operating system and select the appropriate offline installer. Check if you are using a 32bit or a 64bit operating system and make sure you select the correct version of the installer.

2. FastQC for quality control (http://www.bioinformatics.babraham.ac.uk/projects/fastqc/). Unpack and make the wrapper script fastqc on the top-level executable via:

```
$ chmod 755 fastqc
```

This requires a suitable JRE.
3. Trimmomatic for quality and adapter trimming (http://www.usadellab.org/cms/?page=trimmomatic) [18]. Unpack and it is ready to use. Requires a suitable JRE.
4. BSMAP for alignment (https://code.google.com/archive/p/bsmap/) [19]. Install as you usually would under Linux: First, unpack the source code:

```
$ tar xzvf bsmap-2.90.tgz
```

Change into the unpacked directory and make executable binary:

```
$ cd bsmap-2.90/
$ make
```

Install the binary into system default path:

```
$ make install
```

5. Picard tools for duplicate removal (http://broadinstitute.github.io/picard). Again, just unpack and it is ready to use. Requires a suitable JRE.
6. SAMtools for SAM/BAM manipulation [20]. (https://github.com/samtools/samtools) Click on "Releases" and download, preferably, the latest release. Again, install as you usually would under Linux: First, unpack the source code:

```
$ tar xjvf samtools-1.7.tar.bz2
```

Change into the unpacked directory and make executable binary:

```
$ cd samtools-1.7/
```

```
$ make
```

Install the binary into system default path:

```
$ make install
```

7. DSS for identification of differentially methylated regions (https://bioconductor.org/packages/release/bioc/html/DSS.html) [21]. Install by opening R and typing

```
# try http:// if https:// URLs are not supported
source("https://bioconductor.org/biocLite.R")
biocLite("DSS")
```

3 Methods

3.1 Library Preparation

Library preparation is carried out with a number of commercially available kits. These kits are subject to continuous development and contain detailed instruction manuals that are tailored to the respective kit version. As such, we do not provide step-by-step protocols, but rather focus on critical steps, and provide Notes for troubleshooting. We recommend the use of a paired-end sequencing protocol with 150 nt read length, as it currently provides the maximum yield and the highest cost-efficiency.

1. Genomic DNA isolation—Isolate genomic DNA from >10 mg of tissue using the MasterPureTM DNA Purification Kit (Epicentre) or a similar, column-based purification kit, following the manufacturer's instructions. This procedure should result in >2 μg of genomic DNA (*see* **Notes 1** and **2**), which should be eluted in water.
2. Fragmentation of genomic DNA—Fragmentation of DNA is performed using a Covaris fragmentation system, such as M220 or S2. The overall fragment size distribution should lie between 50 and 550 bp, with a peak target size at 150–200 bp, also depending on the desired read and insert size length (*see* **Note 3**).
3. Library preparation

   Library preparation is done using Illumina's TruSeq PCRfree Library Prep Kit (LT).

(a) End repair: Following Illumina's protocol (TruSeq DNA PCR-Free Library Prep Reference Guide 15036187 D), perform end repair to generate blunt-ended fragments.

(b) Size selection and cleanup: We recommend using a bead-based size selection and cleanup that only excludes large fragments using 1.6× beads, following the manufacturer's instructions. If an additional removal of small fragments is desired, first 1.6× beads, and then 0.3× beads are used as described in the Illumina protocol.

(c) Adenylation of 3′ ends: A-tailing is done following the manufacturer's protocol to prevent fragments from ligating to each other during the subsequent adapter ligation.

(d) Adapter ligation: adapters have a corresponding "T" overhang to ligate the adapter to the fragment. The ligation is performed following the manufacturer's protocol.

(e) Double bead cleanup: (1.6× beads) following the manufacturer's protocol. When using multiplexed libraries, a careful cleanup of remaining adapters is paramount due to possible index hopping on some Illumina platforms (*see* **Note 4**).

4. Bisulfite treatment

   Bisulfite treatment is done using the Epitect Kit (Qiagen). Bisulfite treatment can be performed following the Illumina protocol (Whole Genome Bisulfite Sequencing for Methylation Analysis, Illumina, Part #15021861, Rev. B). To ensure maximum retrieval of DNA after treatment, substitute **step 12** (elution in buffer) by eluting with water. The first eluate should be used for two additional elution steps.

5. Cleanup of bisulfite-converted DNA

   Cleanup of bisulfite-converted DNA is done following the Illumina protocol (Whole Genome Bisulfite Sequencing for Methylation Analysis, Illumina, Part #15021861, Rev. B).

6. Amplification of library

   The amplification is done using Kapa HiFi HotStart Uracil+ ReadyMix (2X). Perform the PCR following the protocol on the technical data sheet from the Kapa kit. Standard primers can be used. We recommend the following primers:

   -P5 (Illumina Forward Primer): 5'AATGATACGGCGACCACCGA 3'

   -PCR Primer PE2.0 short: 5' CAAGCAGAAGACGGCATACGAGAT 3'

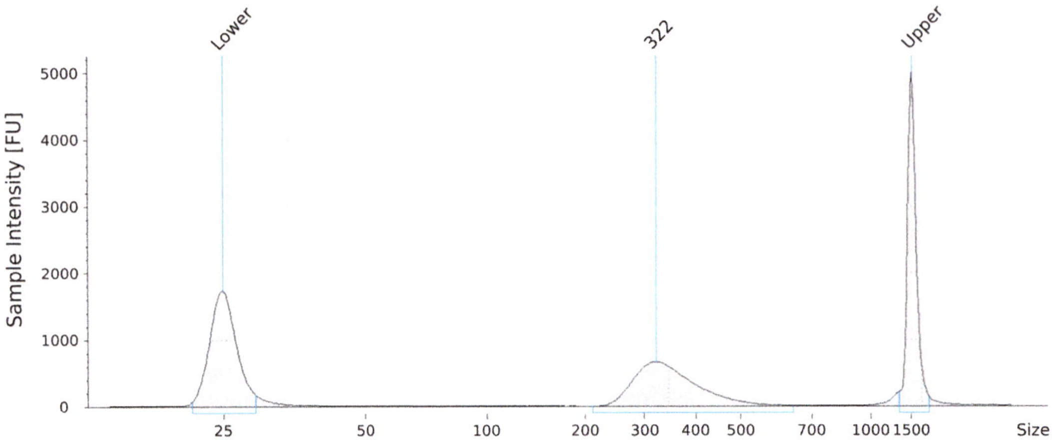

Fig. 1 TapeStation gel electrophoresis of a WGBS library. Visible peaks include the lower and the upper marker (as indicated) and the library with a peak size of 322 bp. The library peak is clearly defined and the fragment size corresponds to the desired insert size. Flat peaks or multiple peaks can be an indication of adapter contamination or hairpins

7. Bead cleanup

   Perform bead cleanup (1× beads) following the Illumina TruSeq PCRfree Library Prep Kit (LT) protocol. The elution volume should be 30 μL.

8. Quality control

   Library quality control is done using Qubit and Bioanalyzer/TapeStation. Again, a single peak is desired and the overall distribution should not be too flat; *see* Fig. 1. If the library aliquots from **step 4**. look good individually, pool and repeat library quality control. Depending on the sequencing platform, PhiX DNA is added to the library to improve library complexity, as suggested by the respective protocols from Illumina.

3.2 Bioinformatics

1. *Data features*—DNA bisulfite sequencing typically results in millions of short-read sequences (usually 100–150 bp), which are stored in FASTQ formatted files. Due to the bisulfite conversion of unmethylated cytosine residues, the reads show a low number of cytosines. To confidently determine the correct methylation status of a given cytosine position, it is necessary to sequence to sufficiently high coverage—10× is currently considered a good target coverage for meaningful analyses. Single- and paired-end sequencing is possible. However, paired-end sequencing is preferred since it contains spatial information and can increase mapping rates in low complexity regions, where each mate is an anchor for the other. Generally, mapping efficiencies are reduced to 70–80% for bisulfite-converted reads: spontaneous degradation can occur due to the aggressive bisulfite treatment, and the reduced complexity of the C-depleted reads further reduces the efficiency. To estimate bisulfite conversion rates, unmethylated spike-in DNA

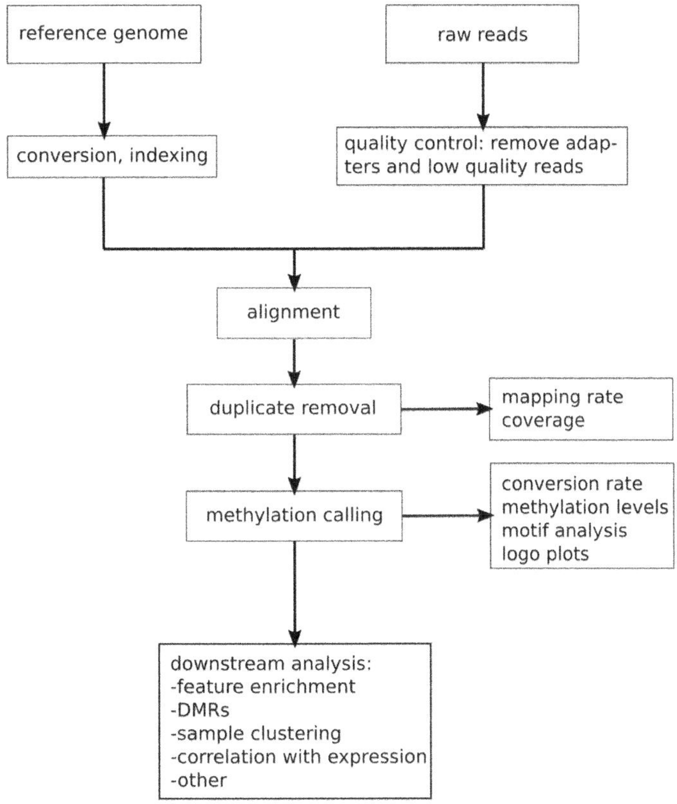

Fig. 2 Flowchart of the WGBS bioinformatics pipeline. Standard procedures include quality control, quality trimming, alignment to a reference, removal of duplicates, and methylation calling. Multiple metrics can be called from the mapping, such as mapping rate, overall coverage, and percentage of duplicates. After methylation calling, commonly used metrics include the estimation of the conversion rate, global methylation levels, motif analysis, and logo plots. Downstream analyses can include methylation feature enrichment, identifying differentially methylated regions, clustering of samples, or correlation with expression data

can be included in the sequencing. Additional details for this procedure are provided in Chapter 12 [17]. Especially when studying differential methylation, it is advisable to analyze biological replicates to account for natural variance and increase the significance of the results. All given commands here assume paired-end sequencing data, but can easily be adjusted for single-end data. An overview of a bioinformatics pipeline is provided in Fig. 2.

2. *Quality control*—Performing quality controls of sequencing data has become a standard protocol before any further downstream processing. Identifying problems early allows necessary preprocessing steps to correct mistakes before false

interpretations can be made, and helps reduce biases in the data. FastQC is the standard tool used to perform such a quality control of raw sequencing reads. It extracts different quality metrics such as low base calling quality, levels of sequence duplication, adapter contamination and overrepresented sequences, and low conversion rate (e.g., C to T <0.9). Many of these problems can be addressed as described below and during mapping to reduce their impact on the analysis. The most important of these metrics are described below. Also *see* Chapter 12 [17] for a complete description of the metrics produced by FastQC.

```
$ fastqc sample1.fastq.gz sample2.fastq.gz
  ...
```

(a) "Per base sequence quality": The per base sequence quality plot shows the base-call quality score levels along read positions in a FASTQ file. The quality score Q is a Phred score which is defined as $Q = -10 * \log_{10} P$, where P is the base-calling error probability. Scores of >30 are generally considered good quality, since the probability for an incorrect base call in this position is <0.001. A decrease in quality toward the 3' end of the read is common for Illumina platforms, as well as an overall lower quality of mate 2 in paired-end sequencing, and trimming should be performed accordingly. *See* Fig. 3a for an example.

(b) "Per base sequence content": The plot shows the base composition for each read position along the read length for all reads contained in the FASTQ file. It should show an even distribution for each base along the read. In the case of bisulfite sequencing, it is also an indicator for successful bisulfite conversion. Bisulfite treatment should convert all unmethylated cytosines to thymines, while all methylated cytosines remain the same. Depending on the species and their overall methylation levels, the proportion of Cs is often <1 % and is called the non-conversion rate. Accordingly, thymine levels will appear elevated in WGBS. It is possible to observe an increase of cytosines toward the 3' end, which can be due to decreased sequencing quality and/or adapter contamination. This issue can be resolved as described in 3 (*Data preprocessing*). *See* Fig. 3b for an example.

(c) "Sequence duplication levels": high levels of duplication, i.e., reads that are exact copies of each other, or map to the exact same positions in the reference genome, are most

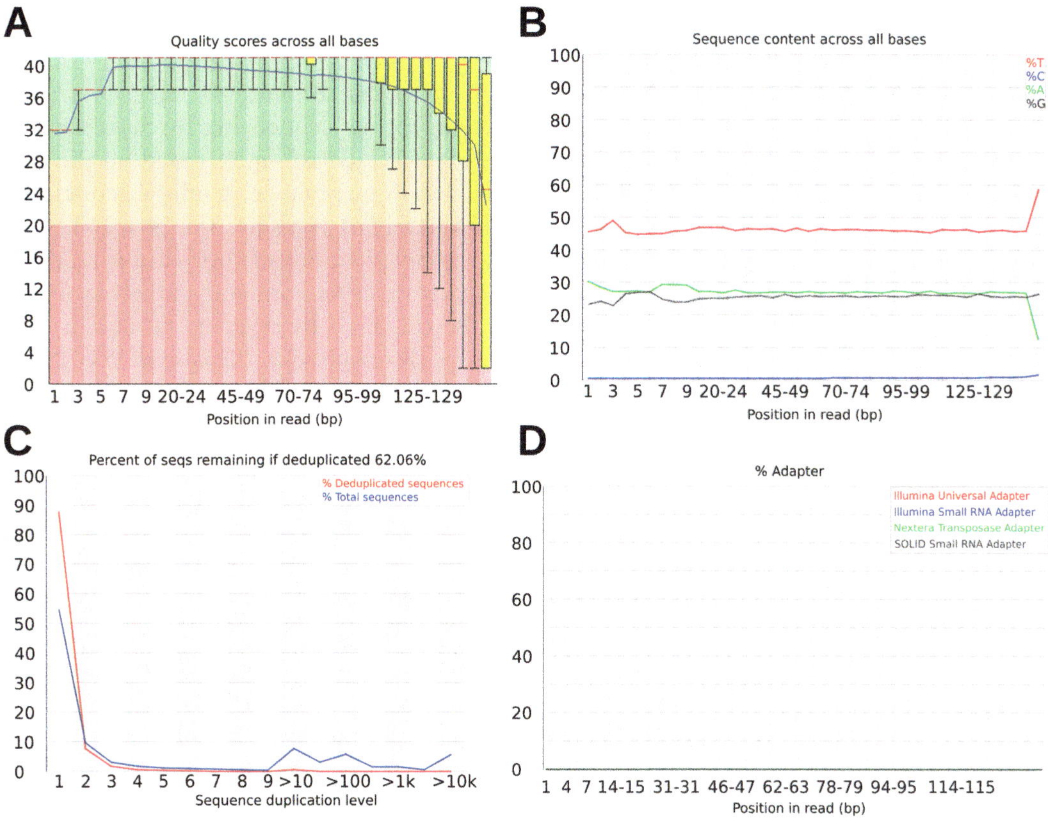

Fig. 3 FASTQC plots. (**a**) per base sequence quality should be high along the read position, and usually drops toward the end. It is generally lower for Mate 2 in paired-end sequencing. (**b**) The base composition per read position should show an even distribution along the read. Cytosine levels will be highly depleted in WGBS, while thymine levels will be elevated. Rising or falling levels toward the end of the read typically result from lower base qualities toward the end of the read. (**c**) Sequence duplication levels are shown for total and de-duplicated sequences. Duplication removal is always recommended before further downstream analyses. (**d**) Plot showing undetectable levels of adapter content and overrepresented sequences

commonly caused by PCR errors or by fragmentation bias during library preparation. They are an indication of low complexity of the sequencing library, and may lead to incorrect quantification when being included in downstream processing. Therefore, they should be removed or ignored for further methylation quantification analyses. *See* Fig. 3c for an example.

(d) "Adapter content and overrepresented sequences": These metrics are useful for detecting adapter contamination. If overrepresented sequences occur in high percentages, or if the plot shows high adapter content along positions in the read (typically at the 3' end), it is a sign of adapter contamination which should be removed before any further analysis. Again, this is described in 3. (*Data preprocessing*). *See* Fig. 3d for an example.

(e) "Kmer content": The kmer content will detect overrepresented sequences in regular sequencing by calculating an observed over expected ratio of kmers based on the base composition in the library. Since bisulfite sequencing reads contain low levels of cytosine, the probability for the presence of multiple cytosines in one read is very low. Therefore, observing cytosine-containing kmers leads to a strong skew in the observed to expected ratio. This plot should therefore be disregarded in the case of WGBS.

Generally, FastQC uses conservative cutoffs, and it is likely to see some of the metrics failed for the initial quality control of raw sequencing reads. In this case, it is advisable to perform the usual preprocessing steps (*see* Subheading 3) and repeat the FastQC analysis.

3. *Data preprocessing*—Before any further processing, adapters should be trimmed from the reads, and low-quality bases or reads should be removed. We suggest using Trimmomatic, a command line tool which can trim adapters, remove leading or trailing low-quality bases, or remove read parts with overall low quality. We first trim Illumina adapters (provided by Trimmomatic; *see* **Note 5**) and clip leading or trailing bases falling below a Phred quality threshold of $Q > 22$ (the ideal threshold to retain maximum information without compromising quality [22]). We then perform a sliding window approach, where for each window the average quality must again not fall below a score of 22. Finally, we only retain reads with a minimum length of 36 bases. For paired-end sequencing, Trimmomatic provides four fastq files, two paired files, one for each mate if both read mates successfully completed the trimming process, and two single files where only one read from a pair was successfully processed. The flag -threads can be included for parallel read trimming.

```
$ java -jar trimmomatic-0.35.jar PE -phred33
sample_R1.fastq.gz sample_R2.fastq.gz
sample_R1_trimmomatic_paired.fastq.gz
sample_R1_trimmomatic_unpaired.fastq.gz
sample_R2_trimmomatic_paired.fastq.gz
sample_R2_trimmomatic_unpaired.fastq.gz
ILLUMINACLIP:TruSeq3-PE.fa:2:30:10 LEADING:22
TRAILING:22 SLIDINGWINDOW:4:22 MINLEN:36
```

4. *Alignment to a reference genome*—Since bisulfite treatment converts unmethylated cytosines to uracils, the overall number

of Cs will be relatively low. Some bisulfite mapping tools therefore require an initial in silico C to T conversion of the reference genome to allow higher mapping efficiency of converted reads (*see* Chapter 12 [17] for an example). In contrast, the widely used bisulfite mapping software BSMAP [19] maps by enumerating all C to T combinations for each *k*seed-length and then aligns by hashing and extension. It is a stand-alone tool and supports a variety of experimental setups and mapping parameters. These include single and paired-end reads, variable numbers of allowed mismatches, quality cutoffs and parallel mapping using up to 8 threads (*see* **Note 6**). BSMAP outputs mapping in SAM/BAM format [19], which is supported by the vast majority of downstream tools. An important metric of the sequencing and alignment is the mapping efficiency, which is provided in the BSMAP output. This output gives information on the number of mapped paired reads, and the number of mapped single reads. Additional statistics can be computed using SAMtools flagstat, including the total number of mapped bases/genome coverage and the number of duplicates (after duplicate marking, *see* below). The reads of the unmethylated spike-in control can be used for calculating the conversion efficiency. Effective deamination results in conversion rates >0.99 (*see* **Note 7**). The reference genome should be closely related to the sequenced sample/individual, since sequence polymorphisms may lead to false methylation calling categories (*see* **Note 8**). The flag -p should be used for parallel computing. Output and error messages can be saved to .log and .err files, respectively.

```
$ bsmap -a sample_R1_trimmomatic_paired.fastq.
gz -b sample_R2_trimmomatic_paired.fastq.gz -d
reference.fa -o sample_alignment.sam
1>sample_alignment.log 2>sample_alignment.err
```

5. *Duplicate removal*—As mentioned before, duplicates are likely to be introduced into the library by PCR and fragmentation bias during library preparation. When included in downstream processing, they can introduce false quantifications and should therefore be ignored or removed. Picard is a set of command line tools designed for the manipulation of high-throughput sequencing data formats. The tool "MarkDuplicates" from the Picard suite finds reads that map to the same coordinates in a reference, and considers these duplicates. By default, it only marks them, and the flag REMOVE_DUPLICATES should be set to "true" to remove them from the BAM (SAM) file. The outputted metric file provides information on the performed duplicate removal. *See* **Note 9** for further details.

```
$ java -jar picard.jar MarkDuplicates
I=sample_alignment.sam REMOVE_DUPLICATES=true
O=sample_alignment_rmdup.bam
M=sample_alignment_rmdup.metric
```

6. *Methylation* calling—After duplicate removal, methylation calling determines the methylation ratio of each covered cytosine position after mapping. BSMAP includes a python script to calculate methylation ratios for each covered cytosine in the reference, along with a number of other metrics concerning methylation. It requires an installation of SAM tools, and the SAM tools location needs to be specified in the command. The methylation table generated by BSMAP contains the following columns:
   (a) chromosome
   (b) coordinate
   (c) strand
   (d) sequence context (2 nt upstream to 2 nt downstream in Watson strand direction)
   (e) methylation ratio
   (f) number of reads covering this locus
   (g) number of unconverted Cs in the reads at this locus

These can be used to compute a number of useful metrics, including the global CpG coverage and global methylation levels as well as analyzing sequence context for methylation.

```
$ python methratio.py -s samtools-location/ -d
reference.fa -p -o sample_alignment_rmdup.ratio
sample_alignment-rmdup.bam 1>sample_rmdup.
ratio.log 2>sample_rmdup.ratio.err
```

7. *Downstream analyses*—A variety of metrics and downstream analyses can be performed once mapping and methylation calling are complete. However, there are no standard tools and much depends on the general structure of the methylation landscape in the respective organism (e.g., sparse methylation, mosaic methylation, ubiquitous methylation). Similarly, much depends on the scope of the study, which can range from the methylation analysis of specific sites (e.g., transcription factor binding sites) to phylo-epigenetic analyses of a large number of species. Below, we provide some examples that can be used as starting points for more detailed analyses.

When working with an annotated reference genome, a feature enrichment can be performed to determine in which parts of the genome methylation occurs (gene body, repeats, etc.). This is especially interesting for insect methylomes, since the methylation target features can be highly diverse. Methylation levels along the genome can be displayed in genome browsers, together with available annotations. This can be performed locally using the Integrative Genome Viewer (http://software.broadinstitute.org/software/igv/home), or remotely using the UCSC Genome Browser (http://genome.ucsc.edu). To examine the relationship between methylation and transcription, methylation levels of gene regions can be correlated with the corresponding gene expression levels. It is advisable to select different types of genic regions (promoter, intragenic, enhancer) to determine which methylation regions have the strongest correlation to transcription.

Comparisons between different datasets from the same organism can provide interesting insight into context-dependent methylation changes (*see* **Note 10**). The identification of differentially methylated cytosines or regions (DMCs or DMRs) between different biological conditions can reveal potential regulatory functions of methylation. DSS [21] is a tool which implements hierarchical modeling to identify differentially methylated regions with high sensitivity. It is available on Bioconductor and it also allows the identification of DMRs between samples for which no replicates are available (*see* **Note 11**). DSS recommends to apply a smoothing step for adjustable window sizes to calculate DMRs. In genomes with very low methylation, this is not advisable as it might obscure genuine differences in methylation at single cytosines.

When comparing WGBS datasets from different species, it is often helpful to visualize methylation landscapes. This can be done by showing representative methylation tracks or by visualizing and comparing methylation levels of sliding windows across entire genomes. For the latter, violin plots or similar can be useful.

Additionally, if RNA-seq data is available, computing the correlation between gene methylation and the corresponding expression levels (FPKM, TPM) can give insight into a potential impact of DNA methylation on gene expression. Computing this correlation can be done using freely available software packages like rsem (https://github.com/deweylab/RSEM). This correlation is particularly interesting for differentially methylated genes or for genes that are associated with differentially methylated regions.

4 Notes

1. Bisulfite deamination requires harsh chemical conditions that result in an unpredictable degree of DNA fragmentation. The

use of sufficient starting amounts of DNA ensures the availability of larger fragments after deamination.

2. DNA should be prepared from fresh-frozen or ethanol-preserved samples. The purity of the sample is important for the efficiency of bisulfite deamination. Library preparation from DNA of formalin-fixed and paraffin-embedded samples is often more problematic.
3. To obtain sufficient genome coverage, approximately 200 million reads (PE150 or PE100) per 1 Gbp of genome sequence should be planned for each sample.
4. Multiple samples can be multiplexed on single lanes. However, index switching can be a problem with particular barcodes and library preparation kits and needs to be minimized [23]. A careful cleanup after adapter ligation is paramount to minimize so-called index hopping. When using adapters on both ends of the reads, index hopping is practically eliminated. Consult Illumina's white paper "Effects of Index Misassignment on Multiplexing and Downstream Analysis" if you plan on sequencing multiplexed libraries.
5. The software Trimmomatic comes with a suite of standard adapters; users can easily add their own adapters to the directory or to the FASTA file provided.
6. If the reference genome is highly fragmented, the performance will become very slow. In this case, it is advisable to combine the fragments into roughly 100 supercontigs, and reassign methylation levels to original contigs after methylation calling.
7. It is advisable to include unmethylated DNA to the sample to determine the efficiency of the bisulfite conversion and estimate the conversion rate. Unmethylated PhiX lambda phage spike-in is a common choice. The deamination rate of spike-in controls should be >99%.
8. DNA should be prepared from a strain that is closely related to the strain that was used for the reference genome sequence (*see* Subheading 3.1, **step 1**). Wild insect populations can have a high degree of genetic variability, which can lead to an artifactual detection of a comparably high level of methylation outside the canonical CpG dinucleotide context. This problem can be solved by the generation of a custom reference genome, e.g., by parallel whole-genome sequencing of the DNA sample.
9. Picard's MarkDuplicates is considered the standard tool for duplicate removal. However, if BAM files are very large, memory limits can occur, even on large memory machines. To overcome this, and enhance performance, pre-sort the BAM file using SAMtools, and set Picard's parameter ASSUME_SORTED to true. Additionally, the parameter

SORTING_COLLECTION_SIZE_RATIO determines the memory footprint used by some of the sorting collections. When running out of memory, this parameter can be reduced from its default of 0.25.

10. Depending on the scope of the analysis and the expected methylation pattern, 2-3 biological replicates should be analyzed to account for biological variability.
11. For any DMR identification software that includes a smoothening step (like DSS), caution is advised for the smoothening parameters. In sparsely methylated genomes, smoothening might obscure differential methylation.

Acknowledgments

We thank Günter Raddatz, Stephan Wolf, Nicolle Diessl, and the DKFZ Genomics and Proteomics Core Facility for helpful discussions.

References

1. Goll MG, Bestor TH (2005) Eukaryotic cytosine methyltransferases. Annu Rev Biochem 74:481–514
2. Suzuki MM, Bird A (2008) DNA methylation landscapes: provocative insights from epigenomics. Nat Rev Genet 9(6):465–476
3. Jones PA (2012) Functions of DNA methylation: islands, start sites, gene bodies and beyond. Nat Rev Genet 13(7):484–492
4. Schubeler D (2015) Function and information content of DNA methylation. Nature 517 (7534):321–326. https://doi.org/10.1038/nature14192
5. Jaenisch R, Bird A (2003) Epigenetic regulation of gene expression: how the genome integrates intrinsic and environmental signals. Nat Genet 33 Suppl:245–254
6. Feng S, Cokus SJ, Zhang X, Chen PY, Bostick M, Goll MG, Hetzel J, Jain J, Strauss SH, Halpern ME, Ukomadu C, Sadler KC, Pradhan S, Pellegrini M, Jacobsen SE (2010) Conservation and divergence of methylation patterning in plants and animals. Proc Natl Acad Sci U S A 107(19):8689–8694
7. Zemach A, McDaniel IE, Silva P, Zilberman D (2010) Genome-wide evolutionary analysis of eukaryotic DNA methylation. Science 328 (5980):916–919
8. Falckenhayn C, Boerjan B, Raddatz G, Frohme M, Schoofs L, Lyko F (2013) Characterization of genome methylation patterns in the desert locust *Schistocerca gregaria*. J Exp Biol 216. (Pt 8:1423–1429. https://doi.org/10.1242/jeb.080754
9. Lyko F, Foret S, Kucharski R, Wolf S, Falckenhayn C, Maleszka R (2010) The honey bee epigenomes: differential methylation of brain DNA in queens and workers. PLoS Biol 8(11):e1000506
10. Raddatz G, Guzzardo PM, Olova N, Fantappie MR, Rampp M, Schaefer M, Reik W, Hannon GJ, Lyko F (2013) Dnmt2-dependent methylomes lack defined DNA methylation patterns. Proc Natl Acad Sci U S A 110(21):8627–8631
11. Falckenhayn C, Carneiro VC, de Mendonca Amarante A, Schmid K, Hanna K, Kang S, Helm M, Dimopoulos G, Fantappie MR, Lyko F (2016) Comprehensive DNA methylation analysis of the Aedes aegypti genome. Sci Rep 6:36444. https://doi.org/10.1038/srep36444
12. Bewick AJ, Vogel KJ, Moore AJ, Schmitz RJ (2017) Evolution of DNA methylation across insects. Mol Biol Evol 34(3):654–665. https://doi.org/10.1093/molbev/msw264
13. Clark SJ, Harrison J, Paul CL, Frommer M (1994) High sensitivity mapping of methylated cytosines. Nucleic Acids Res 22 (15):2990–2997
14. Huang Y, Pastor WA, Shen Y, Tahiliani M, Liu DR, Rao A (2010) The behaviour of 5-hydroxymethylcytosine in bisulfite

sequencing. PLoS One 5(1):e8888. https://doi.org/10.1371/journal.pone.0008888

15. Lister R, Ecker JR (2009) Finding the fifth base: genome-wide sequencing of cytosine methylation. Genome Res 19(6):959–966
16. Lyko F, Maleszka R (2011) Insects as innovative models for functional studies of DNA methylation. Trends Genet 27(4):127–131
17. Asselman J (2018) Bioinformatic analysis of methylation patterns using bisulfite sequencing data. In: Brown S, Pfrender ME (eds) Insect genomics, methods in molecular biology, vol XXX. Springer-Nature, New York
18. Bolger AM, Lohse M, Usadel B (2014) Trimmomatic: a flexible trimmer for Illumina sequence data. Bioinformatics 30 (15):2114–2120. https://doi.org/10.1093/bioinformatics/btu170
19. Xi Y, Li W (2009) BSMAP: whole genome bisulfite sequence MAPping program. BMC Bioinformatics 10:232
20. Li H, Handsaker B, Wysoker A, Fennell T, Ruan J, Homer N, Marth G, Abecasis G, Durbin R, Genome Project Data Processing S (2009) The sequence alignment/map format and SAMtools. Bioinformatics. 25 (16):2078–2079. https://doi.org/10.1093/bioinformatics/btp352
21. Feng H, Conneely KN, Wu H (2014) A Bayesian hierarchical model to detect differentially methylated loci from single nucleotide resolution sequencing data. Nucleic Acids Res 42(8): e69. https://doi.org/10.1093/nar/gku154
22. Del Fabbro C, Scalabrin S, Morgante M, Giorgi FM (2013) An extensive evaluation of read trimming effects on Illumina NGS data analysis. PLoS One 8(12):e85024. https://doi.org/10.1371/journal.pone.0085024
23. Sinha R, Stanley G, Gulati GS, Ezran C, Travaglini KJ, Wei E, Chan CKF, Nabhan AN, Su T, Morganti RM, Conley SD, Chaib H, Red-Horse K, Longaker MT, Snyder MP, Krasnow MA, Weissman IL (2017) Index switching causes "Spreading-Of-Signal" among multiplexed samples in illumina HiSeq 4000 DNA sequencing. BioRxiv. https://doi.org/10.1101/125724

Chapter 12

Bioinformatic Analysis of Methylation Patterns Using Bisulfite Sequencing Data

Jana Asselman

Abstract

Epigenetic factors, including DNA methylation, play a crucial role in the development, behavior, and stress response of insects yet the analysis of DNA methylation patterns remains quite challenging. This chapter will introduce the different technologies for DNA methylation analysis and present a general methodology for the analysis of DNA methylation patterns using the commonly used technology of bisulfite sequencing. The chapter will give a short overview of the sequencing technology itself and will primarily focus on presenting the bioinformatic and statistical analysis methodology of bisulfite sequencing data to study DNA methylation patterns.

Key words DNA methylation, Bioinformatics, Bisulfite sequencing

1 Introduction

Over the past decade, it has become clear that epigenomic information in addition to genomic information plays a vital role in many biological processes [1]. Epigenomic information is often referred to as an additional layer of information on top of the DNA strand that does not affect the DNA sequence [1]. The most well-known example of an epigenetic modification is the methylation of DNA through the addition of a methyl group. Until recently, it was thought that DNA methylation in eukaryotes was exclusively found at cytosines, but it is now recognized that DNA methylation can also occur at adenine residues [2–5]. The recent interest in DNA methylation combined with the revolution in sequencing technologies has motivated many researchers to expand their genomic investigations beyond simple characterization of variation in DNA sequences to descriptions of genome-wide methylation patterns. While it is now feasible to include the study of DNA methylation patterns into experimental designs, the detection and analysis of DNA methylation patterns can be challenging given the high number of available technologies and data analysis tools.

Susan J. Brown and Michael E. Pfrender (eds.), *Insect Genomics: Methods and Protocols*, Methods in Molecular Biology, vol. 1858, https://doi.org/10.1007/978-1-4939-8775-7_12,

The current chapter will start with a very brief overview of the different technologies while providing an in-depth introduction and subsequent methodology for the analysis of methylation patterns with one of the most commonly used technologies.

Sequencing technologies to determine DNA methylation patterns evolve rapidly and vary in their applications. The technologies can be grouped into three major classes: technologies using enzyme digestion, technologies using affinity enrichment, and technologies using sodium bisulfite [6]. Additionally, a small subset of technologies exists that do not include sequencing and therefore do not require genomic information. These technologies make use of liquid chromatography coupled with mass spectrometry (LC-MS) or high performance liquid chromatography (HPLC) [7]. These approaches are limited to quantifying total or global methylation levels and do not provide any information on the specific patterns of methylation of genes or genomic regions. Therefore, these technologies will not be discussed further here. Sequencing technologies using enzymatic digestion such as Methyl-Seq make use of restriction enzymes that are methylation sensitive and are thus inhibited by CpG methylation [6]. Affinity enrichment technologies such as MeDIP-Seq use specific antibodies that have an affinity for methylated genomic regions [6]. Bisulfite sequencing technologies use sodium bisulfate to treat the genomic DNA [6]. This chemical reaction results in the deamination of unmethylated cytosines to uracil whereas methylated cytosines remain unaffected and methylation status can be inferred by comparing the genomic differences between the bisulfite-converted DNA and the original reference (Fig. 1). Bisulfite sequencing technologies are most commonly used in arthropod methylation studies and insect methylation studies in particular. Therefore, the current chapter will focus on the analysis of DNA methylation patterns using bisulfite sequencing technologies. In contrast to other technologies, bisulfite sequencing infers methylation status directly from the sequence itself and allows the detection of methylation patterns at single base resolution (Fig. 1). A thorough comparison of all the different technologies for the analysis of DNA methylation patterns can be found in the review of Laird [6].

The bioinformatic analysis of DNA methylation patterns requires a basic understanding of the bisulfite sequencing technology and the potential errors or biases that can occur during the entire protocol. Indeed, good bioinformatic practices need to account for all potential biases and errors that have been introduced during the sequencing protocol and how they may affect the downstream processing and analysis.

Bisulfite sequencing starts from a standard DNA sample and requires no specific extraction procedures, but it is recommended to have clean genomic DNA as residual proteins may infer with the bisulfite conversion [8]. Most protocols require less than 1 μg of

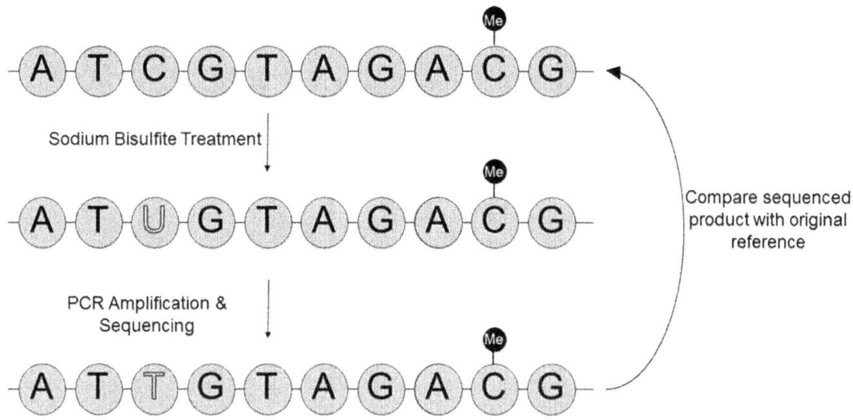

Fig. 1 Bisulfite sequencing consists of sodium bisulfite treatment and subsequent PCR amplification and sequencing of DNA. Me refers to an additional methyl group

input DNA and specific protocols are available for bisulfite sequencing of low input samples such as embryonic cells [9]. The standard bisulfite sequencing protocols start with the shearing or enzymatic digestion of the DNA followed by a denaturation of the DNA and a sodium bisulfite treatment [8]. These steps are followed by purification of the bisulfite-converted DNA and subsequent PCR amplification and sequencing. Currently, a wide set of commercial kits are available from various manufacturers and have been evaluated for their use in different experimental designs [10]. A fully detailed protocol on DNA extraction, library preparation, and sequencing for whole genome bisulfite sequencing is provided in Chapter 11 [11].

The most common errors or biases in the bisulfite sequencing protocol are bisulfite conversion errors and PCR bias [6, 8]. The bisulfite conversion errors are the result of incomplete conversion of unmethylated cytosines to uracil. This incomplete conversion will lead to the increase in methylation levels as unconverted unmethylated cytosines will be incorrectly assessed as methylated cytosines. To account for this upward bias in methylation levels due to incomplete bisulfite conversion, unmethylated DNA such as λ phage DNA will be spiked in the sample. The bisulfite conversion rate can then be determined through bioinformatics by comparing the detected methylation level of the spiked unmethylated DNA with the theoretical level of zero methylation. The exact procedure of determining the bisulfite conversion error will be discussed in detail in Subheading 3.4. In contrast, the incorrect conversion of methylated cytosines to uracil, i.e., false negatives, is negligible and almost nonexistent when following standard protocols [8]. PCR bias during the sequencing of bisulfite-converted DNA occurs because methylated and unmethylated molecules have different sequences after conversion as PCR is sequence- and strand-specific

[12]. This bias may lead to an inaccurate estimation of methylation levels. PCR bias is usually controlled at the primer level and by controlling the parameters of the PCR reaction [12]. In addition to PCR bias, a common issue is PCR duplication. This issue is not limited to bisulfite sequencing but to next-generation sequencing technologies in general. PCR duplication refers to the increased amplification levels for a particular sequence or sequences in the library compared to the general amplification levels across all sequences. PCR duplication can be detected during the quality control of the data and can be removed during the bioinformatic analysis.

The next sections will discuss the bioinformatic tools available, the specific statistical methods to analyze DNA methylation patterns and considerations that need to be taken into account, and finally discuss the most common pitfalls during the analysis. The chapter will not provide an exhaustive list of all potential tools and programs but rather focus on the most commonly used and user-friendly programs for researchers starting with DNA methylation analysis [13]. An alternative method specifically tailored to whole genome bisulfite sequencing is described in Chapter 11 [11]. Furthermore, bioinformatic tools evolve quickly and new tools develop at a rapid space. However, the current chapter will highlight the major principles of DNA methylation analysis regardless of the bioinformatic tool or software program used.

2 Materials

2.1 Data Input Requirements

The bioinformatics analysis protocol described here in detail is tailored to Illumina sequencing data. Other bisulfite sequencing data can be used on the condition that the data files are fastq files. Both single-end and paired-end sequencing data can be used. However, as mentioned in Chapter 11, paired-end sequencing will result in a higher mapping efficiency and is particularly recommended for bisulfite sequencing. There are no specific requirements for the data analysis for insert size and read length, although some recommendations can be found in Chapter 11 [11]. Both directional and nondirectional libraries can be analyzed. The methods can also be used to analyze reduced representation bisulfite sequencing data. This does require specific options for certain steps and is described in detail where necessary in the sections below.

2.2 Hardware Requirements

Sequencing data files are generally large files in terms of a couple of gigabytes per file that require powerful desktops and preferably high performance computing (hpc) infrastructure. In addition, the majority of bioinformatic tools and software programs can sometimes be quite challenging to install as they perform best in a unix environment. Therefore, the use of hpc infrastructure, which is often readily available nowadays at many universities, also comes with the advantage that many bioinformatic programs and tools are already installed or can be installed by the computer experts managing the hpc infrastructure.

When you do not have access to hpc infrastructure, the easiest option is often to install biolinux on a powerful desktop [14]. Biolinux is an open-source bioinformatics workstation that comes with more than 250 bioinformatic tools and packages preinstalled. This option is particularly recommended for users with no expertise in command line software installation and compilation, and in modifying environmental paths as the majority of the required software and tools are preinstalled. If you prefer working in a windows environment, an additional compiler such as MinGW or CygWin will be necessary while in a mac environment Xcode is recommended.

The bioinformatic tools and programs needed recommend a system with at least five cores and more than 16 Gb of RAM.

2.3 List of Bioinformatic Tools

2.3.1 FastQC

FastQC is a simple and easy-to-use program for quality control. It is freely available and can be downloaded at https://www.bioinformatics.babraham.ac.uk/projects/fastqc/. It is written in Java and one of the few tools that can be installed on any operating system without any problems. It can be run in a graphical application or via the command line. This program is preinstalled on Biolinux.

2.3.2 Trim Galore—Trimmomatic—Cutadapt

Trim Galore and Trimmomatic are tools for adapter and quality trimming. Both are publically and freely available. The first is a wrapper script for FastQC and cutadapt (https://www.bioinformatics.babraham.ac.uk/projects/trim galore/). This program requires a functional installation of cutadapt, which in turn is dependent on python and a C compiler. Trimmomatic is a flexible read aligner and does not have any dependencies [15] (*see* **Notes 1** and **2**).

2.3.3 Bismark—Bowtie2

Bismark is a program to map bisulfite sequenced read to a reference genome and perform methylation calls [16]. For the mapping step, Bismark makes use of bowtie2 which is a short-read aligner commonly used for aligning DNA and RNA reads [17]. Bismark and Bowtie require a compiler to run properly on windows or mac systems. Bowtie2 is preinstalled on Biolinux, (*see* **Notes 1** and **2**).

2.3.4 R—Bioconductor

R is a free software for statistical computing and can be easily installed via https://www.r-project.org/. Bioconductor is an open-source software for bioinformatics and provides a comprehensive set of analysis tools for sequencing data that can be used within the R environment [18]. Both are preinstalled on Biolinux.

2.3.5 Seqmonk

Seqmonk is a tool to visualize and analyze sequencing data, including bisulfite sequencing data. It includes options for statistical analysis and it is publically and freely available at https://www.bioinformatics.babraham.ac.uk/projects/seqmonk/.

3 Methods

This section is divided into four subsections detailing the consecutive steps in the analysis of bisulfite sequencing from quality control to trimming and mapping and concluding with the statistical analysis [13]. Each of the subsections makes use of different bioinformatic programs and tools and is therefore discussed separately. While the general succession of steps is similar to other types of DNA sequence data, specific issues can arise when dealing with bisulfite sequencing data and these will be discussed in the following subsections.

3.1 Quality Control of Bisulfite Sequencing Data with FastQC (see Notes 3, 4 and 5)

Bioinformatic analysis starts with quality control of the obtained sequence data. This step is crucial to identify potential errors or contamination that may have occurred during the experiment or the sequencing and to make sure you are working with good quality data. Therefore, all the sequencing files can be loaded into the FastQC program for quality control analysis. FastQC supports most major file extensions including fastq, compressed fastq, bam, and sam format. For users unfamiliar with the command line, the FastQC program can be run from the graphical user interface by simply opening the data files one by one. A tutorial video is available online (https://www.youtube.com/watch?v=bz93ReOv87Y). Alternatively, the FastQC program can also be executed in the command line with a single command to analyze multiple files, while the help command provides an overview of all other possible options:

```
C:\users\fastqc> fastqc file1.fastq file2.fastq
C:\users\fastqc> fastqc --help
```

In the first command, the user can specify all file names for which a fastQC report will be generated. This will result in an html report and a zip folder containing all the figures. The second command allows the user to view all help information available about fastQC in the command line.

The FastQC output report contains 12 modules and is available in html format. Examples of reports with different contamination types, good and bad quality data, are available online for consultation and comparison, provided by the developers of the program (https://www.bioinformatics.babraham.ac.uk/projects/fastqc/). For each of the modules, the program will generate a green (good), orange (warning), or red (bad) flag. Orange or red flags do not necessarily mean bad quality data or a failed experiment but require the user to carefully examine the data and treat it accordingly by, for example, removing contamination or trimming low quality bases. Here, a brief overview will be given of each of the 12 modules and how to interpret them. The reader is also referred to Chapter 11 for a detailed description of the most important modules in FastQC. The tutorial video and the examples referred to above can be used for a more detailed description of the different sections. The interpretation of the FastQC results will determine how the data will be trimmed and which sequences will need to be removed in Subheading 3.2.

1. Module 1 on basis statistics gives a brief summary of the data which includes the type of encoding for the sequencing file, number of sequences, GC percentage.
2. Module 2 reports the per base sequence quality at each position using a BoxWhisker plot using Phred scale scores. In general, a Phred score of at least 30 at each position is recommended, which represents 99.9% accuracy of the base call, and which will be the case for most sequencing data. A common issue in bisulfite sequencing data can be a decrease in Phred scale score with increasing base position, particularly toward the end of the read.
   - *Potential trimming action (Subheading 3.2): Remove all bases with a Phred score lower than 30 at the 3' end of the read.*
3. Module 3 reports the overall quality score for each sequence; again users can expect that the majority of the sequences will have a quality score of 30 or higher. Sequences with lower scores can indicate a quality loss within the run.
   - *Potential trimming action (Subheading 3.2): remove the read if the average quality is below 30.*
4. Module 4 depicts the per base sequence content, which is expected to be similar across all bases in a random library. For bisulfite sequencing data, this module will always result in a red flag. Indeed, in the case of bisulfite sequencing data, the library contains almost no cytosines as all unmethylated cytosines have been converted to thymines and the proportion of methylated cytosines is usually fairly low. Thus the proportion of cytosines will be a line close to 0% while the proportion of thymines will result in a line close to 50%. When the levels for all bases are between 20% and 30%, it usually indicates that the bisulfite conversion was not successful.

5. Module 5 reports the GC content across the entire sequence compared to a modeled normal GC distribution. Again, given the bisulfite conversion of unmethylated cytosines to thymines, this section will result in a red flag for bisulfite sequencing data as there are few cytosines present in the data.
6. Module 6 reports the per base N content, which refers to the proportion of base calls reported as N. When a base call is reported as N, it means that the sequencer is unable to make a base call with sufficient confidence and report an N to the user. This should be either a very low proportion or zero in good quality sequencing data.
7. Module 7 reports the distribution of sequence lengths, which is in most cases should be uniform.
8. Module 8 reports the amount of duplicated sequences. As mentioned in the introduction, bisulfite sequencing, as well as many other sequencing technologies, can suffer from PCR duplicates. Bismark (Subheading 3.3) always removes PCR duplicates in the downstream analysis. In general, this should be a small proportion of the reads. However, when many duplicates are present, this proportion will increase.
9. Module 9 reports potential overrepresented sequences and potential sources. A warning or red flag in this section can indicate adaptor contamination or other sources of contamination. It could also be the result of a highly duplicated region in the genome of the sample studied. This issue can be easily confirmed by using a BLAST algorithm to match the overrepresented sequence to the genome in question or to a general database such as NCBI to identify the source or origin of the sequence. Contaminated sequences can be filtered out in the downstream analysis, but too much contamination leads to less data or less sequences than expected and may affect the results.
10. Module 10 reports potential adapter contamination. Adapter contamination, when present, can be removed in the downstream analysis.
    - *Potential trimming action (Subheading 3.2): adapter trimming.*
11. Module 11 screens for K-mer content, which is short exactly duplicated sequences. This section can give warnings or red flags when libraries with random primers were used as there is often an incomplete sampling of the genome with random primers and therefore a bias in k-mer content. This issue is visible in the first 5–10 base positions and usually will not affect the downstream analysis.
    - *Potential trimming action (Subheading 3.2): trim the first 5–10 base positions.*

12. Module 12 reports the per tile sequence quality. This report is particularly useful to detect a loss of quality that is associated with a particular part of flow cell in case of Illumina sequencing data. Warnings or red flags in this section suggest technical errors and issues that are inherent to the sequencer and the sequencing process.

The analysis and interpretation of FastQC data will result in a decision on the data quality:

- The data is not of sufficient quality due to issues with the samples or sequencing (e.g., failed or very low bisulfite conversion, low sequence quality for the majority of the bases across the entire read length).
- The data is of sufficient quality to continue with the next step in the analysis.

3.2 Trimming of the Data with Trim Galore or Trimmomatic (see Notes 3, 4 and 5)

In most cases, trimming of the data is recommend to remove small sources of adapter contamination or low quality bases or sequences as they can interfere with the downstream analysis. Trimming can be done with a variety of different programs. Here, the two most popular ones will be discussed, Trim Galore and Trimmomatic. The programs primarily differ in specific available options. Depending on the data, one may be preferred over the other. For whole genome bisulfite sequencing, both tools perform quite well and the choice is usually dependent on the user's preference. For reduced bisulfite sequencing data, Trim Galore has specific options to process this type of data, which is not the case with Trimmomatic. In addition, Trim Galore will automatically detect Illumina adapters and options can be added for other adapters while Trimmomatic requires you to specify and pick the correct file. Both programs need to be run through the command line as no graphical interface is available. While these programs can still be executed on a standard desktop, it is recommended to use hpc infrastructure or a very powerful desktop.

Regardless of the program used, it is important to note that the order in which the commands are specified is the order in which they will be executed. Therefore, it is always crucial to first trim adaptor sequences as identifying adaptors after part of the read is already trimmed is more challenging. Both programs use raw fastq files as input. These input files can be zipped or gzipped.

Trim Galore will execute the following commands consecutively:

1. Find and remove adapter sequences. If no specific adapter sequence is specified, it will default to general Illumina adapters. The function *--adapter* can be used to specify specific adapter sequences. Alternative options are *–Illumina*, *--Nextera*, for Illumina and Nextera adapter sequences.

2. Trim low quality reads from the 3′ end with the function *–q*.
3. Remove short reads with the option *–length*.

For reduced representation bisulfite sequencing libraries, the additional option *–rrbs* can be added. For paired-end libraries, the additional option *–paired* can be added. Additionally, the phred scale quality scores can be changed from phred33 (default) to phred64 by adding the option *–phred64*. The full manual can be found at https://github.com/FelixKrueger/TrimGalore/tree/master/Docs. An example command is illustrated below that will run trim galore for paired-end libraries using standard adapter removal settings, generate a fastqc report at the end, trim low quality reads from the 3′ with a threshold of 30, and use a phred64 scale.

```
C:\users\mysequencingfiles> perl trim_galore --
paired -fastqc -q 30 --phred64 sequence_paired1.
fq.gz sequence_paired2.fq.gz
```

Trimmomatic will execute the following commands consecutively, if specified consecutively in the command line:

1. Find and remove adapter sequencing with the function *ILLUMINACLIP*. This function allows specifying a fasta file with the adapters, followed by the number of allowed mismatches, the threshold for the accuracy of the match between two adapter ligated reads, and the threshold for the accuracy of the match between any adapter sequence and a read.
2. Remove bases at the 3′ end and 5′ end that are below a predefined quality threshold, using the functions *TRAILING* and *LEADING*, respectively.
3. Remove short reads with the option *MINLEN*. Reads below the specified minimum length will then be removed.

For paired-end libraries the option *PE* is specified. Phred scale quality scores can again be specified using the options *–phred64* or *–phred33*. The full manual can be found at http://www.usadellab.org/cms/?page=trimmomatic. An example command is illustrated below that will run trimmomatic on paired-end libraries with a phred scale of 64, remove Illumina adapters, remove bases at the 5′ and 3′ end with a quality below 3, and remove reads with a length lower than 36. Additional information on Trimmomatic can also be found in Chapter 11 [11].

```
C:\users\mysequencingfiles> java -jar trimmomatic.jar PE
-phred64 sequence_paired1.fq.gz sequence_paired2.fq.gz
```

```
output_forward_unpaired.fq.gz output_reverse_paired.fq.
gz output_reverse_unpaired.fq.gz ILLUMINACLIP:adapters.
fa:2:30:10 LEADING:3 TRAILING:3 MINLEN:36
```

The new trimmed fastq files can now be used for mapping and analysis of bisulfite sequencing data. It is, however, recommended that the quality control described in Subheading 3.1 is repeated to ensure all trimming commands were executed properly.

3.3 Mapping and Analysis of the Data Using Bismark (see Notes 3, 4 and 5)

The most important step in the analysis of bisulfite sequencing data is the mapping of the sequencing data to a genome reference and the identification of methylated bases, i.e., methylation calls. Bismark will be used to map the data and report methylation calls. Other tools are available such as BSMap discussed in Chapter 11 [11]. We suggest Bismark here, given that it provides a good combination of processing speed, genomic coverage, and quantitative accuracy as well as a full comprehensive pipeline that does not require any additional tools. Furthermore, it has specific options for reduced representation bisulfite sequencing data. We refer the reader to [19] for a comprehension discussion of the most common mappers for bisulfite sequencing data. Bismark depends on a functional version of bowtie2 and is therefore best executed on a powerful desktop or hpc infrastructure. A high-quality genomic reference that is closely related to organism, preferably the genomic reference of the genotype of interest, is crucial in the mapping and analysis of the data. Indeed, bisulfite sequencing data contains primarily three bases A, T, and G and very low levels of C, particularly in arthropods. Methylation calls can be identified at single base position as they will have a C in the sequence data and a C in the genome reference at that exact position, while unmethylated bases will have a T in the sequence data and a C in the genome reference. Detecting methylation status in this fashion implies that care should be taken when using a genome reference of organism that differs from the organism of interest since any C/T single nucleotide polymorphisms could be incorrectly called as unmethylated cytosines.

1. Prior to any analysis in Bismark, an index of the genome reference needs to be built for the mapping process. Rather than specifying the specific filename, specify the folder name that contains the genome reference. Bismark makes use of bowtie2 for indexing and this needs to be specified in the command line.

```
C:\users\mysequencingfiles>
bismark_genome_preparation --bowtie2 --verbose
Genome_Reference
```

2. Trimmed reads from Subheading 3.2 can now be mapped to the genome reference. This is a time-consuming process and is best conduced on hpc infrastructure. Again, since Bismark makes use of bowtie2 for mapping, this needs to be specified in the command line. Additional options can be added and follow the many options available via bowtie2. The option --*bam* will generate a bam output file. Next, the folder containing the index of the genome reference needs to be specified, followed by the names of the files containing trimmed reads. In case of paired reads, they are both specified using *-1* and *-2* for the forward and reverse read, respectively. Then additional bowtie2 options can be added. Options include a minimum fragment length for paired end alignments, in which the paired end alignment consists of the two reads and the gap; this can be specified using *–I*, with a default of zero. In addition, *-X* will allow the user to specify a maximum fragment length for paired alignments in the same manner, with a default of 500. When obtaining low overall alignment, modifying these settings could potentially improve the alignment rate. The larger the difference between I and X, the slower bowtie2 will run. Additionally, improvements can be made by specifying a function for the minimum alignment score using *--score_min*. The scoring function in bowtie2 is f(x)=a+b*x in which x is the read length and a and b are parameters. The option *-score_min* allows you to specify a and b as follows: L,a,b. Phred scale scores can also be specified. More information on scoring functions and default settings can be found in the bowtie2 manual http://bowtie-bio.sourceforge.net/bowtie2/manual.shtml. The output of the file is in standard BAM format and can be incorporated in most genome browsers for visualization. An example of running bismark can be found below that will execute bismark and bowtie2 for a paired-end dataset (data type is fastq, hence the option -q) with a minimum fragment length of 200 and a maximum fragment length of 1000 while specifying a specific scoring function.

```
C:\users\mysequencingfiles> bismark --bowtie2
--bam Reference/ -1 Trimmed_data_read1.fq -2
Trimmed_data_read2.fq -I 200 -X 1000 -q --
score_min L,0,-0.6 --phred33
```

3. After mapping, it is recommended that users run the deduplication function in bismark to remove potential PCR duplicates for whole genome bisulfite sequencing data. It is not recommended to implement this function for reduced representation bisulfite sequencing data. Overall, the amount of PCR duplicates should be very low in high-quality data.

```
C:\users\mysequencingfiles>
deduplicate_bismark --bam
Trimmed_data_bismark_bt2_pe.bam
```

4. The next step is extracting the methylation calls. Methylation calls are specified in different contexts: CpG, CHG, and CHH. CpG context refers to cytosine methylation in which the methylated cytosine is directly followed by guanine. CHG and CHH refer to cytosine methylation in which the methylated cytosine is not directly followed by a guanine as H refers to another nucleotide than guanine. Bismark uses the letters z, x, and h to refer to a cytosine in CpG, CHG, and CHH context, respectively. An uppercase letter refers to a methylated cytosine while a lowercase letter refers to an unmethylated cytosine. When the methylation context cannot be inferred correctly, the letter u is used. By default, bismark will generate output files for methylation calls for top and bottom strands, resulting in 6 files (1 for each strand and 1 for each methylation context) per sample. Alternative options include merging top and bottom strands in the output files (*--comprehensive*), providing a bedgraph format as output file (*--bedgraph*), and generation of a genome-wide cytosine report (*--cytosine_report*).

```
C:\users\mysequencingfiles> bismark_methylation_extrac-
tor --gzip Trimmed_data_bismark_bt2_pe.deduplicated.bam
```

A more detailed overview of all specific options for each Bismark function can be found online https://rawgit.com/FelixKrueger/Bismark/master/Docs/Bismark_User_Guide.html.

3.4 Statistical Analysis of Bisulfite Sequencing Data

Once methylation calls are obtained, the major research questions often include identifying significant differences in methylation between samples. However, prior to any statistical analysis, accounting for a potential incomplete bisulfite conversion is crucial. Indeed, as described in Subheading 1, bisulfite conversion efficiency is rarely 100%. As a result, some unmethylated cytosine will not be converted to thymines. To account for this error, unmethylated lambda phage DNA can be used. This DNA is usually spiked in the samples prior to bisulfite conversion and sequencing. In theory, all cytosines in the lambda phage DNA sequences should be converted to thymines as they are all unmethylated. Therefore, determining the methylation level of lambda phage DNA gives a good estimate of the bisulfite conversion efficiency and can be used in a binomial testing procedure to identify true positives or true methylation calls [20–22].

1. Determining the bisulfite conversion efficiency: repeat **steps 3.2** and **3.3** using the lambda phage DNA sequence or the DNA sequence of the unmethylated reference as a reference. The output will report the methylation level of the unmethylated reference in the sample, for which the reported value is the false positive rate.
2. Account for false positives: based on the bisulfite conversion efficiency, true positives can be identified using a binomial testing strategy in which the probability of success equaling the false positive rate and the number of trials n is the number of reads for each cytosine. This test can be done in any statistical program of choice. When using R, the function *binom.test* can be used.

After determining the true methylated cytosines, different statistical options are available depending on the sample type and research question. One can study differential methylation at base level or at a higher functional (e.g., gene level) or regional level (e.g., regions of 100 or more base pairs). While studying differential methylation at the base level can be interesting, most datasets will not result in significant findings due to the high number of statistical tests and the multiple testing correction. Therefore, it is often more interesting to group cytosines into higher organizational levels such as genes or CpG islands and reduce the number of statistical tests. This can be done either by grouping bases according to a region of x number of bases or by using a genomic feature file and grouping based on genes, introns, exons, or other genomic features.

Regardless of analyzing the data at base level or higher, the data can be analyzed as normalized levels or ratios by using continuous value statistics or by using counts of methylated versus non-methylated reads using count statistics. Count statistics include Fisher's exact tests and Chi-Square tests. These are generally not recommended as they operate on single replicates and thus do not take advantage of sample replication or account for sample variation. In addition, these basic count statistics do not account for differences in coverage. Normalized levels generally are defined by the ratio of methylated reads versus non-methylated or total reads. These values can then be used in standard continuous test statistics including ANOVA and *t*-test. While these tests can account for replicates easily, the power is significantly reduced. In addition, tiered approaches can be considered in which differentially methylated genes or regions are first identified using continuous test statistics leading to a smaller set of cytosines to be tested at the individual base level and hence a less stringent multiple testing correction. Ultimately, the choice of test statistics should depend on the experiment, the research question, and the user's informed decision on the benefits and limitations of the different methods.

Statistical approaches can be implemented by the user in any statistical environment including R and matlab. Yet, several packages and tools are available that have developed efficient functions for the statistical analysis of methylation data. Below, the most common packages and tools for statistical analysis are listed, including the functions available for methylation analysis in those tools:

- Seqmonk: This is a graphical analysis package that allows different methods of analysis including a fixed window analysis, Chi-square statistics, and logistic regression options. Seqmonk requires bedgraph files as input, which can be directly generated from bismark methylation extractor as described in Subheading 3.3. Analysis of methylation data in SeqMonk is through a graphical interface in which the user can select the statistic of choice, specify regions or features of interest as well as different conditions across samples. More information can be found here: https://www.bioinformatics.babraham.ac.uk/projects/seqmonk/.
- MethylKit is a package developed for R [23] that uses logistic regression or Fisher's exact test to identify differential methylation. The input files can be coverage files or cytosine reports generated directly from bismark methylation extractor as described in Subheading 3.3. The function *methRead* is available to read in multiple files while the function *calculateDiffMeth* will identify differential methylation. The package also allows the user to cluster samples, include a principal component analysis, and report on descriptive methylation statistics. More information can be found here: https://bioconductor.org/packages/release/bioc/html/methylKit.html.
- Bsseq is a package developed for R [24] that uses a Bsmooth algorithm to obtain methylation estimates which can then be used for differential methylation analysis. Data input consists of scaffold, position, number of methylated reads, and coverage. This format will require some preprocessing of the bismark output files to adhere to these input settings. An advantage of this algorithm is that it does not require data from all biological replicates in all treatments for each cytosine. This feature is particularly useful when one or two biological replicates have zero coverage by chance for a specific position and this position would be excluded from the analysis otherwise. Differential methylation is detected using a t-statistic. The function *makeBSseqData* can be used to create a BSobject, which serves as input for the *BSmooth* function and the subsequent *BSmooth.tstat*. *BSmooth* will perform the smoothing and parameter optimization might be necessary depending on the samples. Additionally, the function *dmrFinder* is available to identify differentially methylated region on the results of the t-statistics. More

information can be found here: https://bioconductor.org/packages/release/bioc/html/bsseq.html.

- DSS is a package developed for R [25–27]. In the first steps of the differential methylation analysis, the package relies on the bsseq package described above. Therefore, the input files are the same and the function *makeBSseqData* can be used to create a BSobject which serves as input for the subsequent differential methylation analysis. Smoothing is not necessary but recommended to obtain better estimates for CpG methylation in the case of dense data (e.g., whole genome bisulfite sequencing data); the package makes use of the function *DMLtest* that will first estimate mean methylation levels for each CpG site, estimate dispersions for each CpG site, and then conduct a Wald test to detect differential methylation. While the default null hypothesis is a methylation difference of zero, users can specify a specific threshold as well. In addition, differentially methylated regions can be identified using the function *callDMR*. The major advantage of the DSS package is that it also has a function *DMLfit.multiFactor* that fits a linear model on arcsine transformed methylated data through ordinary least square to the data allowing the user to test more complex designs including potential interactions between treatments or conditions. The test statistics for this model can be obtained using the function *DMLtest.multiFactor*, which uses the Wald test. More information can be found here: http://bioconductor.org/packages/release/bioc/html/DSS.html.

4 Notes

Many errors can occur when running bioinformatic programs graphically or through the command line. Quite a number of errors are common to all tools. When running into an error message, consider the following steps:

1. Is the program installed correctly and working properly? To this end, many example files and datasets are available online. If the error persists, the issue is usually with the installation, the environmental path, or location of the data files. If the error does not persist, the issue is with the data files.
2. Is the environmental path specified correctly? When working on a personal desktop from the command line, the user specifies the program name. However, this program name should be added to the environmental path for the computer to be able to find it. Alternatively, the user can run the command line from the folder location which contains the program files.

3. Are the file locations specified properly? Most errors occur because the computer or hpc cannot find the sequencing files. Double check whether the path to the files is specified properly in the command line and whether the file names are specified correctly. In particular, when working within an hpc infrastructure this is important as depending on the setup of the infrastructure, jobs are not automatically submitted from the same directory as the data files.
4. Given the larger size of the data files, they can become corrupted during subsequent downloading, uploading, and transferring. Therefore, it is always recommended to check the md5sum for the original file and the copied, downloaded, or uploaded file. They should be the same for both files. If not, an error occurred and the file is not properly transferred.
5. Use of the correct syntax. Commands are often case sensitive and the correct syntax is essential. Errors usually occur when using lowercase instead of uppercase letters or vice versa, the use of - instead of --.

In the following subsections, the most common errors will be highlighted and addressed. In general, the error message itself will give a good indication on the error and how to resolve the potential issue.

Notes on Quality Control and FastQC

The most common error with FastQC is an incompatible Java version and can be solved by installing the proper version of Java.

Notes on Trimming with Trim Galore or Trimmomatic

The most common error with Trim Galore is a functional installation of cutadapt and properly referring to the cutadapt file location when running the program. When users are unsuccessful in adding cutadapt to the environmental path of the computer, an alternative solution is to specify the full path containing the location of cutadapt in the command line.

If, after trimming, the fastQC results did not result in the expected improvement or trimming, the most common error is the order of the commands. Most likely, the order was not correctly specified (e.g., specifying trimming reads prior to adapter contamination may result in an incomplete removal of adapter sequences). Rerunning the program with the commands in the correct order (*see* Subheading 3.2) will solve the problem.

Notes on Analysis of Bisulfite Sequencing Data with Bismark

The most common error with Bismark is a functional installation of bowtie2 and properly referring to the bowtie2 file location when running the program. This issue can be resolved by either adding bowtie2 correctly to the environmental path or specifying the full path of the bowtie2 file location in the command line.

Notes on Statistical Analysis of Methylation data

When using any of the statistical packages in R, it is recommended to verify whether the packages and their dependencies work with the current version of R on the computer. If not, the version of R should be changed. Additionally, installation errors of packages or errors in loading libraries usually occur because the necessary dependencies are not installed. This issue can be easily resolved by identifying the dependency that is missing from the error message and install it.

The different packages in R listed above require specific input file types that need to be adhered to strictly, often including column names. Most common errors therefore include incorrect input file types or columns and can be easily changed by generating the correct input file either from bismark directly or modifying the input file in R itself so it adheres to the input requirements.

References

1. Suzuki MM, Bird A (2008) DNA methylation landscapes: provocative insights from epigenomics. Nat Rev Genet 9(6):465–476. https://doi.org/10.1038/nrg2341
2. Fu Y, Luo GZ, Chen K, Deng X, Yu M, Han D, Hao Z, Liu J, Lu X, Dore LC, Weng X, Ji Q, Mets L, He C (2015) N6-methyldeoxyadenosine marks active transcription start sites in Chlamydomonas. Cell 161(4):879–892. https://doi.org/10.1016/j.cell.2015.04.010
3. Greer EL, Blanco MA, Gu L, Sendinc E, Liu J, Aristizabal-Corrales D, Hsu CH, Aravind L, He C, Shi Y (2015) DNA methylation on N6-Adenine in C. elegans. Cell 161(4):868–878. https://doi.org/10.1016/j.cell.2015.04.005
4. Sun Q, Huang S, Wang X, Zhu Y, Chen Z, Chen D (2015) N6-methyladenine functions as a potential epigenetic mark in eukaryotes. Bioessays 37(11):1155–1162. https://doi.org/10.1002/bies.201500076
5. Zhang G, Huang H, Liu D, Cheng Y, Liu X, Zhang W, Yin R, Zhang D, Zhang P, Liu J, Li C, Liu B, Luo Y, Zhu Y, Zhang N, He S, He C, Wang H, Chen D (2015) N6-methyladenine DNA modification in Drosophila. Cell 161(4):893–906. https://doi.org/10.1016/j.cell.2015.04.018
6. Laird PW (2010) Principles and challenges of genomewide DNA methylation analysis. Nat Rev Genet 11(3):191–203. https://doi.org/10.1038/nrg2732
7. Kurdyukov S, Bullock M (2016) DNA methylation analysis: choosing the right method. Biology (Basel) 5(1). https://doi.org/10.3390/biology5010003
8. Warnecke PM, Stirzaker C, Song J, Grunau C, Melki JR, Clark SJ (2002) Identification and resolution of artifacts in bisulfite sequencing. Methods 27(2):101–107
9. Millar DS, Warnecke PM, Melki JR, Clark SJ (2002) Methylation sequencing from limiting DNA: embryonic, fixed, and microdissected cells. Methods 27(2):108–113
10. Holmes EE, Jung M, Meller S, Leisse A, Sailer V, Zech J, Mengdehl M, Garbe LA, Uhl B, Kristiansen G, Dietrich D (2014) Performance evaluation of kits for bisulfite-conversion of DNA from tissues, cell lines, FFPE tissues, aspirates, lavages, effusions, plasma, serum, and urine. PLoS One 9(4): e93933. https://doi.org/10.1371/journal.pone.0093933
11. Gatzmann F, Lyko F (2018) Whole-genome bisulfite sequencing for the methylation analysis of insect genomes. In: Brown S, Pfrender M (eds) Insect genomics, methods in molecular biology, vol XXX. Springer-Nature, New York
12. Warnecke PM, Stirzaker C, Melki JR, Millar DS, Paul CL, Clark SJ (1997) Detection and measurement of PCR bias in quantitative methylation analysis of bisulphite-treated DNA. Nucleic Acids Res 25(21):4422–4426
13. Krueger F, Kreck B, Franke A, Andrews SR (2012) DNA methylome analysis using short bisulfite sequencing data. Nat Methods 9(2):145–151. https://doi.org/10.1038/nmeth.1828
14. Field D, Tiwari B, Booth T, Houten S, Swan D, Bertrand N, Thurston M (2006) Open software for biologists: from famine to feast. Nat Biotechnol 24(7):801–803. https://doi.org/10.1038/nbt0706-801

15. Bolger AM, Lohse M, Usadel B (2014) Trimmomatic: a flexible trimmer for Illumina sequence data. Bioinformatics 30 (15):2114–2120. https://doi.org/10.1093/bioinformatics/btu170
16. Krueger F, Andrews SR (2011) Bismark: a flexible aligner and methylation caller for Bisulfite-Seq applications. Bioinformatics 27 (11):1571–1572. https://doi.org/10.1093/bioinformatics/btr167
17. Langmead B, Trapnell C, Pop M, Salzberg SL (2009) Ultrafast and memory-efficient alignment of short DNA sequences to the human genome. Genome Biol 10(3):R25. https://doi.org/10.1186/gb-2009-10-3-r25
18. Gentleman RC, Carey VJ, Bates DM, Bolstad B, Dettling M, Dudoit S, Ellis B, Gautier L, Ge Y, Gentry J, Hornik K, Hothorn T, Huber W, Iacus S, Irizarry R, Leisch F, Li C, Maechler M, Rossini AJ, Sawitzki G, Smith C, Smyth G, Tierney L, Yang JY, Zhang J (2004) Bioconductor: open software development for computational biology and bioinformatics. Genome Biol 5(10): R80. https://doi.org/10.1186/gb-2004-5-10-r80
19. Kunde-Ramamoorthy G, Coarfa C, Laritsky E, Kessler NJ, Harris RA, Xu M, Chen R, Shen L, Milosavljevic A, Waterland RA (2014) Comparison and quantitative verification of mapping algorithms for whole-genome bisulfite sequencing. Nucleic Acids Res 42(6):e43. https://doi.org/10.1093/nar/gkt1325
20. Bonasio R, Li Q, Lian J, Mutti NS, Jin L, Zhao H, Zhang P, Wen P, Xiang H, Ding Y, Jin Z, Shen SS, Wang Z, Wang W, Wang J, Berger SL, Liebig J, Zhang G, Reinberg D (2012) Genome-wide and caste-specific DNA methylomes of the ants Camponotus floridanus and Harpegnathos saltator. Curr Biol 22 (19):1755–1764. https://doi.org/10.1016/j.cub.2012.07.042
21. Pegoraro M, Bafna A, Davies NJ, Shuker DM, Tauber E (2016) DNA methylation changes induced by long and short photoperiods in Nasonia. Genome Res 26(2):203–210. https://doi.org/10.1101/gr.196204.115
22. Rehan SM, Glastad KM, Lawson SP, Hunt BG (2016) The Genome and Methylome of a Subsocial Small Carpenter Bee, Ceratina calcarata. Genome Biol Evol 8(5):1401–1410. https://doi.org/10.1093/gbe/evw079
23. Akalin A, Kormaksson M, Li S, Garrett-Bakelman FE, Figueroa ME, Melnick A, Mason CE (2012) methylKit: a comprehensive R package for the analysis of genome-wide DNA methylation profiles. Genome Biol 13 (10):R87. https://doi.org/10.1186/gb-2012-13-10-r87
24. Hansen KD, Langmead B, Irizarry RA (2012) BSmooth: from whole genome bisulfite sequencing reads to differentially methylated regions. Genome Biol 13(10):R83. https://doi.org/10.1186/gb-2012-13-10-r83
25. Feng H, Conneely KN, Wu H (2014) A Bayesian hierarchical model to detect differentially methylated loci from single nucleotide resolution sequencing data. Nucleic Acids Res 42(8): e69. https://doi.org/10.1093/nar/gku154
26. Wu H, Xu T, Feng H, Chen L, Li B, Yao B, Qin Z, Jin P, Conneely KN (2015) Detection of differentially methylated regions from whole-genome bisulfite sequencing data without replicates. Nucleic Acids Res 43(21):e141. https://doi.org/10.1093/nar/gkv715
27. Park Y, Wu H (2016) Differential methylation analysis for BS-seq data under general experimental design. Bioinformatics 32 (10):1446–1453. https://doi.org/10.1093/bioinformatics/btw026

Chapter 13

Physical Genome Mapping Using Fluorescence In Situ Hybridization with Mosquito Chromosomes

Maria V. Sharakhova, Gleb N. Artemov, Vladimir A. Timoshevskiy, and Igor V. Sharakhov

Abstract

The development of genomic resources and tools is an important step in designing novel approaches to genetic control of mosquitoes. Physical genome maps enhance the quality of the genome assemblies, improve gene annotation, and provide a better framework for comparative and population genomics studies in mosquitoes. In this chapter, we describe protocols for an important procedure in physical genome mapping—fluorescence in situ hybridization (FISH). We provide details on (1) dissection of salivary glands, ovaries, and imaginal discs for obtaining high-quality polytene or mitotic chromosome preparations; (2) DNA-labeling procedures and extraction of repetitive DNA fractions; and (3) approaches to FISH on polytene and mitotic chromosomes.

Key words Mosquitoes, Fluorescence in situ hybridization, Polytene and mitotic chromosomes, Physical genome mapping

1 Introduction

Mosquitoes from Culicidae family are vectors of various pathogens of human infectious diseases including malaria, dengue, yellow fever, Zika, West Nile fever, and filariasis [1]. The development of genomic resources and tools including physical genome maps is an important step in designing novel approaches to genetic control of mosquitoes. Drawn and photo chromosomal maps have been constructed for about 50 species of the genus *Anopheles* [2]. The physical genome maps based on polytene chromosomes have been developed for six vectors of human malaria including *An. gambiae* [3] (Fig. 1), *An. stephensi* [4, 5], *An. sinensis* [6], *An. atroparvus* [7], and *An. funestus* [8]. A remarkably high-coverage genome map that includes 98.2% of the genome assembly was developed for a Neotropical vector of malaria, *An. albimanus* [9]. High-quality idiograms based on mitotic chromosomes from imaginal discs of fourth instar larvae have been developed for the vector

Susan J. Brown and Michael E. Pfrender (eds.), *Insect Genomics: Methods and Protocols*, Methods in Molecular Biology, vol. 1858, https://doi.org/10.1007/978-1-4939-8775-7_13,

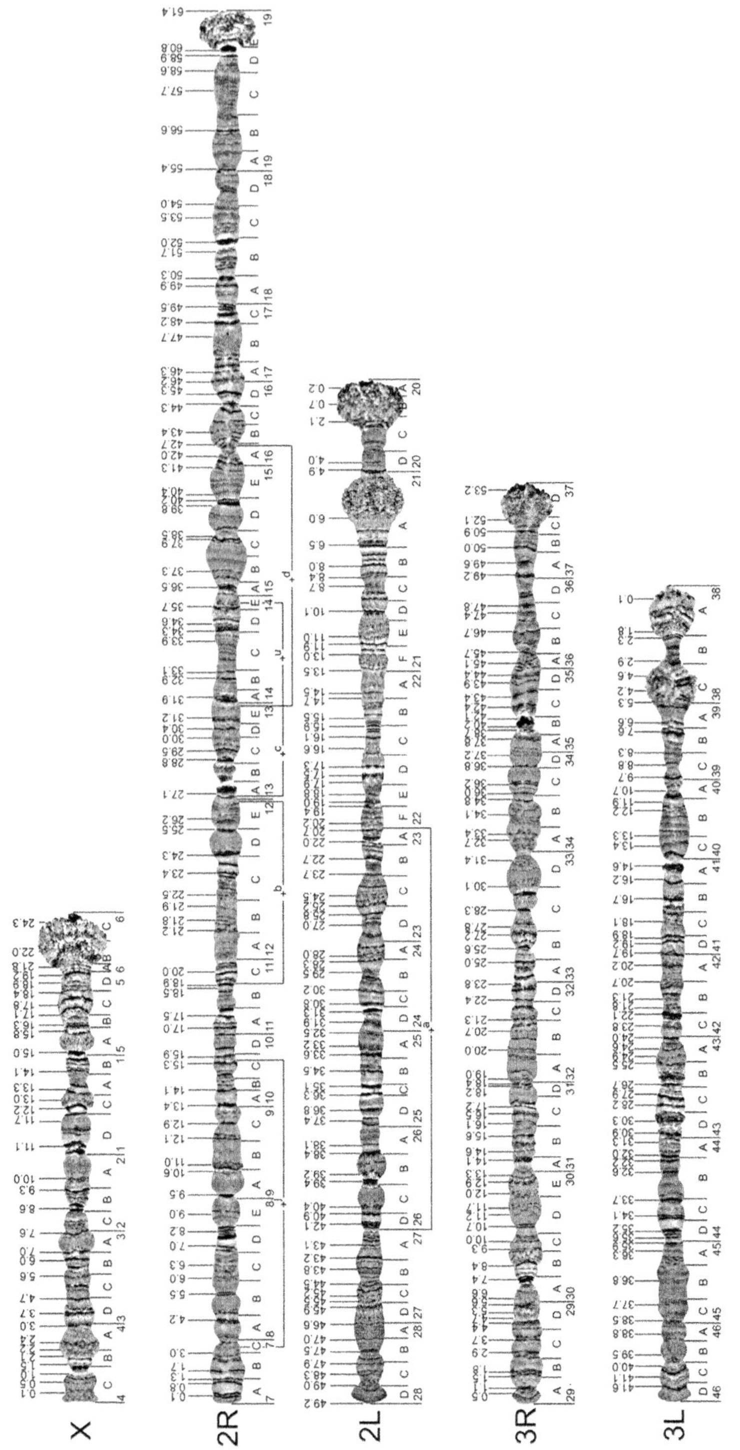

Fig. 1 A cytogenetic photomap of *An. gambiae* ovarian nurse cell chromosomes [3]. Genome coordinates are shown above X, 2R, 2L, 3R, and 3L chromosomal arms. Divisions and subdivisions are indicated by numbers and letters below the chromosomes. The location of common polymorphic inversions $+^{j}$, $+^{b}$, $+^{c}$, $+^{u}$, $+^{d}$, and $+^{a}$ is shown in their standard orientation by brackets below the chromosomes

of yellow fever, Zika, and dengue *Aedes aegypti* [10, 11] and for the vector of West Nile fever and filariasis *Culex quinquefasciatus* [12]. An application of FISH to mitotic chromosomes of *Ae. aegypti* placed 45% [13] of this mosquito genome to precise chromosomal positions (Fig. 2).

In this chapter, we describe protocols for FISH that represent an important procedure for genome mapping in mosquitoes. These protocols include: (1) dissection of salivary glands, ovaries, and imaginal discs for obtaining high-quality polytene or mitotic chromosome preparations; (2) DNA-labeling procedures; extractions of repetitive DNA fractions that are required to block unspecific hybridization of BAC DNA clones; and (3) different approaches to FISH procedures on polytene and mitotic chromosomes. These protocols are modified and updated versions of the previously published protocols [14–16].

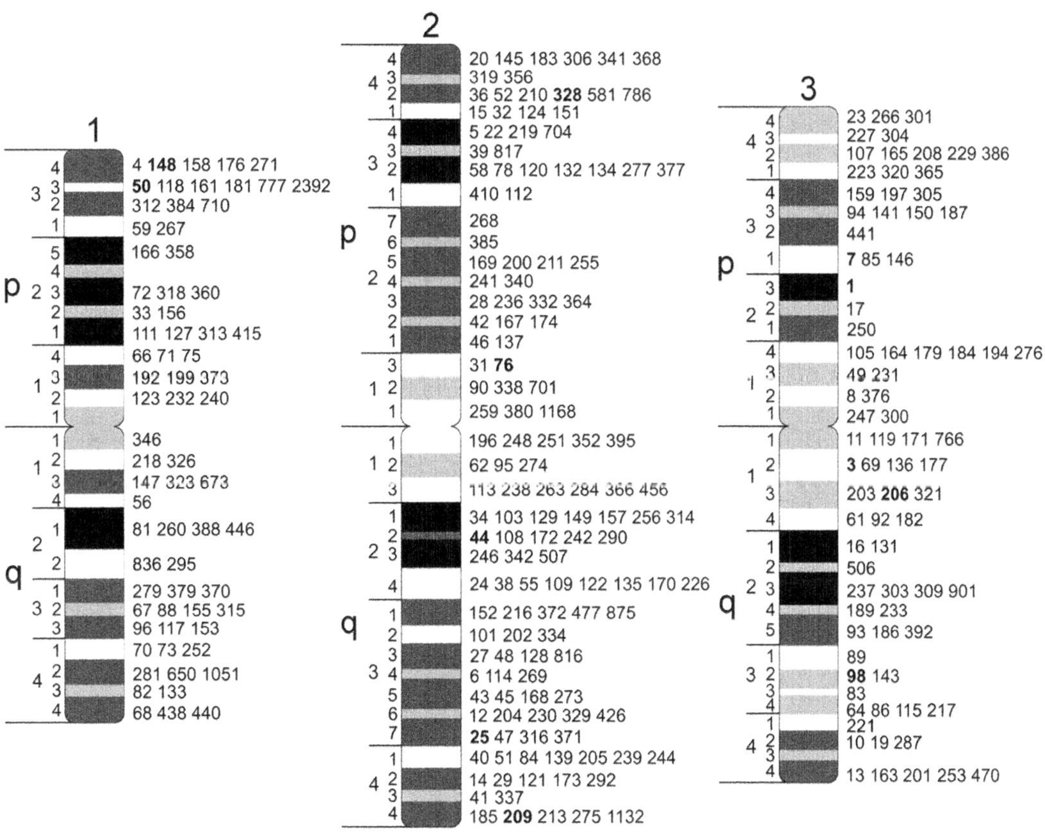

Fig. 2 A physical map of the *Ae. aegypti* genome [13]. Chromosome divisions and subdivisions are indicated on the left side of the idiograms. The positions of supercontigs are shown on the right side of the idiograms. Supercontig names are indicated by the last 2–4 digits. Supercontigs in conflict with previous genetic mapping/genome assemblies are in bold

2 Materials

2.1 Chromosome Preparations

1. 18 × 18 mm microscope coverslips (Fisher Scientific, Cat. No. 12-553-402).
2. 22 × 22 mm microscope coverslips (Fisher Scientific, Cat. No. 12-544-10).
3. Dissecting needles (Fine Science Tools, Cat. No. 10130-10).
4. Dissecting scissors (Fine Science Tools, Cat. No. 15000-03).
5. Olympus CX41 Phase Microscope (Olympus, Cat. No. CX41RF-5).
6. MZ6 Leica stereomicroscope (Leica, Cat. No. VA-OM-E194-354).
7. Needle holders (Fine Science Tools, Cat. No. 26018-17).
8. ThermoBrite™ Slide Denaturation/Hybridization System (Abbott Molecular, Cat. No. 30-144110).
9. Sodium citrate dihydrate (Fisher Scientific, Cat. No. S279-500).
10. Potassium chloride (Fisher Scientific, Cat. No. BP366-500).
11. Propionic acid (Sigma-Aldrich, Cat. No. 402907).
12. Alcohol 200 Proof (Decon Laboratories, Cat. No. 2701).
13. Methanol (Fisher Scientific, Cat. No. A412-4).
14. Acetic acid (Fisher Scientific, Cat. No. A491-212).
15. 20× SSC buffer (Invitrogen, Cat. No. AM9765).

2.2 Probe Preparation

2.2.1 DNA Labeling

1. Qiagen Large Construct Kit (Qiagen, Cat. No. 12462).
2. QIAquick® PCR purification Kit Qiagen, Cat. No. 28104).
3. Cy3-dUTP, Cy5-dUTP (Enzo®, ENZ-42501, ENZ-42502).
4. dATP, dCTP, dGTP, dTTP (Fermentas, Cat. No. R0141, R0151, R0161, R0171).
5. Random Primed DNA Labeling System (Invitrogen, Cat. No. 18187-013).
6. DNA polymerase I (Fermentas, Cat. No. EP0041).
7. DNase I (Fermentas, Cat. No. EN0521).
8. BSA (Sigma-Aldrich, Cat. No. A3294).
9. Sodium acetate trihydrate (Fisher Scientific, Cat. No. BP334-500).
10. Alcohol 200 Proof (Decon Laboratories, Cat. No. 2701).
11. EDTA (Fisher Scientific, Cat. No. S331-500).

2.2.2 Extraction of Repetitive DNA Fractions

1. Qiagen Blood and Cell Culture Maxi Kit (Qiagen, Cat. No. 13362).
2. S1 Nuclease (Fermentas, Cat. No. EN0321).
3. 20× SSC buffer (Invitrogen, Cat. No. AM9765).
4. Alcohol 200 Proof (Decon Laboratories, Cat. No. 2701).
5. Tris base (Fisher Scientific, Cat. No. BP152-1).
6. EDTA (Fisher Scientific, Cat. No. S331-500).
7. Isopropanol (Fisher Scientific, Cat. No. A516-4).
8. Sodium acetate trihydrate (Fisher Scientific, Cat. No. BP334-500).

2.3 FISH

1. ThermoBrite™ Slide Denaturation/Hybridization System (Abbott Molecular, Cat. No. 30-144,110).
2. Olympus BX61 fluorescent microscope (Olympus, Cat. No. BX61).
3. 22 × 22 mm microscope coverslips (Fisher Scientific, Cat. No. 12-544-10).
4. 1 mM YOYO-1 iodide (491/509) solution in DMSO (Invitrogen, Cat. No. Y3601).
5. 10× PBS (Invitrogen, Cat. No. P5493).
6. 10% NBF (neutral-buffered formalin) (Sigma-Aldrich, Cat. No. HT501128).
7. 20× SSC buffer (Invitrogen, Cat. No. AM9765).
8. 99% formamide (Fisher Scientific, BP227500).
9. Acetic acid (Fisher Scientific, Cat. No. A491-212).
10. Alcohol 200 Proof (Decon Laboratories, Cat. No. 2701).
11. Dextran sulfate sodium salt (Sigma-Aldrich, Cat. No. D8906).
12. Hydrochloric acid (Fisher Scientific, Cat. No. A144-500).
13. Nonidet-P40 (NP40) (US Biological, Cat. No. NC9375914).
14. Parafilm (Fisher Scientific, Cat. No. 13-374-10).
15. Pepsin (USB, Cat. No. 9001-75-6).
16. ProLong® Gold antifade reagent (Invitrogen, Cat. No. P36930).
17. Sodium acetate trihydrate (Fisher Scientific, Cat. No. BP334-500).
18. RNase (Sigma-Aldrich, Cat. No. 9001-99-4).
19. Rubber Cement (Fisher Scientific, Cat. No. 50-949-105).
20. Salmon sperm DNA (Sigma-Aldrich, Cat. No. D7656).

3 Methods

3.1 Chromosome Preparations

3.1.1 Polytene Chromosome Preparation from Salivary Glands of Fourth Instar Anopheles Larvae

1. Preserve early fourth instar larvae in Carnoy's Solution: ethanol/glacial acetic in 3:1 ratio and store at −20 °C.
2. Remove one fourth-instar larva from the vial with forceps or needle and place it on a dust-free microscope slide with back upward then put a drop of fresh Carnoy's solution onto it immediately (Fig. 3a) (*see* **Note 1**).

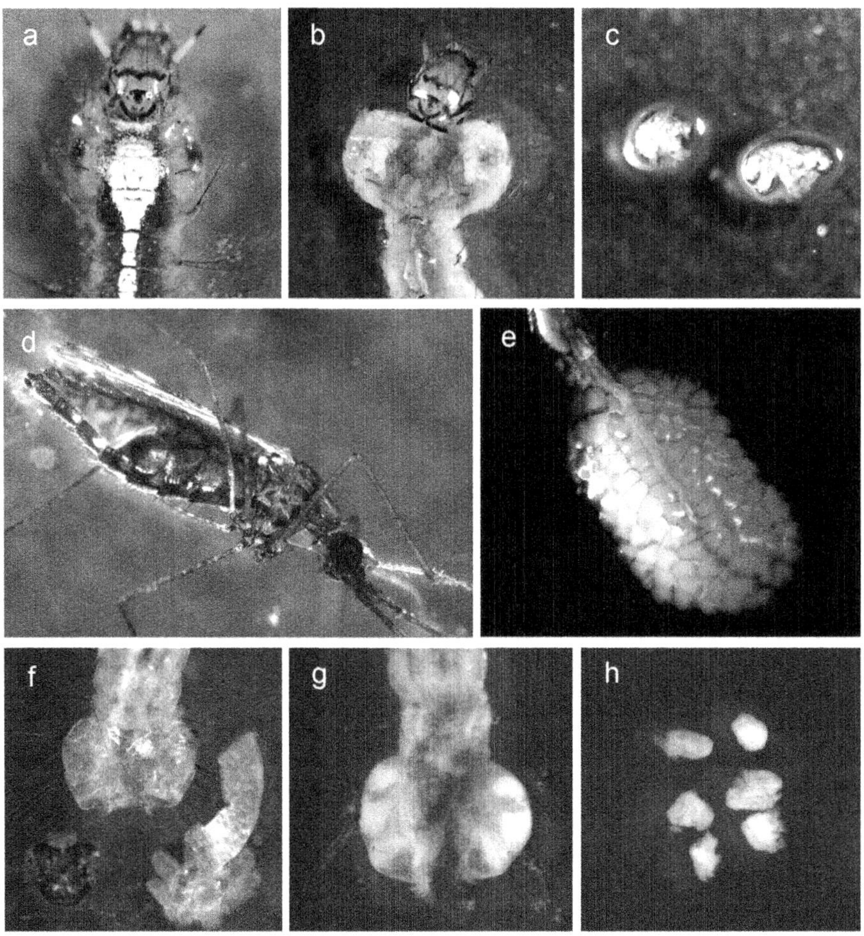

Fig. 3 Mosquito dissections: dissection of salivary glands from fourth instar *An. atroparvus* larvae (**a**–**c**), *An. gambiae* ovaries (**d**, **e**), and *Ae. aegypti* imaginal discs (**f**–**h**). A larva at the fourth instar stage of development is shown on panel **a**. The salivary glands in the opened larval thorax are demonstrated on panel **b**. Panel **c** shows dissected salivary glands. Adult female with ovaries at the Christophers' III stage is indicated on panel **d**. Dissected ovaries at the Christophers' III stage are shown on panel **e**. Panel **f** shows an open thorax, dissected head, and midgut of fourth instar larvae in hypotonic solution. Imaginal discs after application of Carnoy's solution turn white (panel **g**). Dissected imaginal discs are demonstrated on panel **h**

3. Insert a needle from the middle rear of the thorax just underneath the cuticle, and gently move forward to break the thorax cuticle along the mid dorsal line. Carefully open up the thorax and remove the salivary glands from connecting tissue (Fig. 3b). Remove the carcass and other tissue from the slide and put one drop of fresh 50% propionic acid onto the glands (Fig. 3c).
4. Cover glands with a dust-free 18 × 18 mm coverslip and leave them for about 5 min. After 5 min, place a piece of filter paper over the coverslip, hold the four edges with fingers. Gently tap the coverslip with a pencil eraser to release the polytene chromosomes from the salivary gland.
5. Examine the banding pattern and spreading of polytene chromosomes using a phase-contrast microscope.
6. Place slides with good chromosomal preparations in a humid chamber with 4× SSC in the bottom of the chamber, at 60 °C for 15–20 min (*see* **Note 2**). Heating can be done on the Thermobrite machine with the absorbent strips soaked in distilled water. After heating, put slides at 4 °C from several minutes to overnight until immersing the slides into liquid nitrogen.
7. While holding one corner of the slide with forceps or a gloved hand, dip chromosome preparation into liquid nitrogen so that the coverslip is completely immersed. Hold slide in liquid nitrogen until the bubbling stops (usually 10–15 s). Take it out of the liquid nitrogen and immediately remove the coverslip with a razor blade from one corner. Put slide in a slide jar with prechilled 50% ethanol (−20 °C) and keep at 4 °C for at least 2 h. Dehydrate the preparations in slide jar with an ethanol series of 70%, and 90% for 5 min each at 4 °C and then 100% ethanol for 5 min at room temperature (RT). Air-dry, and keep slides in slide box until ready for use in in situ hybridization (*see* **Note 3**).

3.1.2 Polytene Chromosome Preparation from Ovaries from Half-Gravid Anopheles Females

1. Give blood to recently emerged (48–72 h old) females using a live animal or artificial blood feeders, and provide a container with water for them to lay eggs. Container of water should be provided 48 h after blood feeding, and eggs will typically be laid ~72 h after blood feeding. It is important that the mosquitoes lay eggs at least once before feeding them for chromosome preparation. If they do not lay eggs after the first blood feeding, feed them one more time and allow at least one gonotrophic cycle including oviposition.
2. Before the dissection, blood feed mosquitoes again and let them develop ovaries for 25 h at 26 °C and 80% humidity.

3. Prepare fresh modified Carnoy's solution: methanol/glacial acetic acid in 3:1 ratio (*see* **Note 4**). Use only high-quality methanol and glacial acetic acid from trusted brands. Make aliquots of the solution by 0.5–1 mL in 1.5–2 mL tubes.
4. Immobilize mosquitoes by putting a cage into a −20 °C freezer for 1 min. Collect females and place them on ice in a Petri dish before dissection. Keep mosquitoes alive.
5. Choose half-gravid females for dissection. At this stage, the light area occupied by the developing ovaries goes from the sixth dorsal to fourth ventral abdominal segments. The rest of the abdomen looks dark because of blood meal (Fig. 3d).
6. Dissect half-gravid females under a dissection microscope using sharp forceps or needles. Make a cut between third and fourth segments from the end of abdomen. Pull off the three segments with attached ovaries from an abdomen on a dry dust-free slide. Remove all blood and extra tissue because they contain water.
7. Check the stage of ovaries development. They must be at Christophers' III stage [17]. At the underdeveloped stage, follicles are small; a transparent area with nurse cells within follicles has not been developed. At the correct stage, follicles have an oval shape; a transparent area with nurse cells within follicles has a round shape (Fig. 3e). At the overdeveloped stage, follicles have an elongated, slightly bent oval shape (a banana shape).
8. Put ovaries from approximately five females into 500 μL of modified Carnoy's solution in a 1.5 mL Eppendorf tube, and keep them at RT for 24 h. Transfer ovaries to −20 °C for a long-term storage.
9. To make the chromosome preparations, take one ovary out of the vials with forceps (or a transfer pipet), and place it into a drop of Carnoy's solution on microscope slide. After carefully removing tissues, trachea, and blood, quickly separate follicles from one ovary into 2–4 pieces. Depending on the size up to four preparations can be made from one ovary.
10. On 4–8 microscope slides add each piece of the divided ovary to one drop of 50% propionic acid. The number of slides prepared from the ovaries depends on how many pieces each ovary is divided into. Let the pieces of ovary rest in propionic acid for about 5 min until follicles become clear and swell to about twice their original size.
11. For each slide, use a dissecting microscope to separate the cleared follicles from each other and any other tissue or debris on the slide. Remove tissue and debris by wiping it away with a

piece of paper towel, and apply a fresh drop of 50% propionic to the separated follicles.

12. Cover ovaries with a dust-free 22 × 22 coverslip, and leave them for about 5 min. After 5 min, place a piece of filter paper over the coverslip, and hold the four edges with fingers. Gently tap it with a pencil eraser to release polytene chromosomes from the follicles.
13. Examine the banding pattern and spread of polytene chromosomes using a phase-contrast microscope.
14. Place slides with good chromosomal preparations in a humid chamber with 4× SSC in the bottom of the chamber, at 60 °C for 15–20 min. Heating can be done on the Thermobrite machine with the absorbent strips soaked in distilled water. After heating, put slides at 4 °C overnight or until immersing the slides in liquid nitrogen (*see* **Note 5**).
15. While holding one corner of the slide with forceps or a gloved hand, dip chromosome preparation into the liquid nitrogen so that the coverslip is completely immersed. Hold slide in liquid nitrogen until the bubbling stops (usually 10–15 s). Take it out of liquid nitrogen, and immediately remove the coverslip with a razor blade from one corner. Put slide in a slide jar with pre-chilled 50% ethanol (−20 °C), and keep at 4 °C for at least 2 h.
16. Dehydrate the preparations in slide jar with an ethanol series of 50%, 70%, and 90% for 5 min each at 4 °C and then with 100% ethanol for 5 min at RT. Air-dry, and keep slides in slide box until ready for use in FISH (*see* **Note 3**).

3.1.3 Mitotic Chromosome Preparation from the Imaginal Discs of Fourth Instar Larvae of Any Species of Mosquito

1. Hatch mosquito eggs at 28 °C, and after 2–3 days, transfer second or third instar larvae to a chamber kept at a lower temperature. Use 16 °C for *Ae. aegypti* and *Cx. quinquefasciatus* and 22 °C for *An. gambiae*.
2. Place fourth instar larvae on ice for several minutes to immobilize them.
3. Transfer a larva to a slide with a drop of cold hypotonic solution (0.5% sodium citrate dihydrate or 0.075 M potassium chloride), and place it under the stereo microscope.
4. It is important to choose a larva at the correct stage. Select larva with oval imaginal discs for further dissection (Fig. 3f–h). Round or overdeveloped imaginal discs that have been already transformed into legs and wings are not suitable for the chromosomal preparations.
5. Decapitate larva, and cut the cuticle from the ventral side of the larval thorax using dissecting scissors. Make an additional cut in second or third abdominal segment to dissect the gut from the larva.

6. Open the cuticle, and remove the gut and fat body from the larva (Fig. 3f). Remove the hypotonic solution from the slide using filter paper, and add a fresh drop of hypotonic solution directly to the imaginal discs within the larva. Keep the larva in hypotonic solution for 10 min at RT.
7. Remove hypotonic solution using filter paper, and apply Carnoy's solution (ethanol/acetic acid in 3:1 ratio). After adding fixative solution, imaginal discs immediately turn white while the surrounding tissues are light yellow (Fig. 3g) (*see* **Note 1**).
8. Using dissecting needles, remove imaginal discs from the larva, and transfer them to a drop of 50% propionic acid (Fig. 3h). Remove any other tissues, such as the gut and fat body, from the slide. Cover imaginal discs with an unsiliconized 22 × 22 cover slip, and keep at RT for 5–10 min.
9. Cover the slide with filter paper, and squash the tissue by tapping the eraser of a pencil on the perimeter of the cover slip.
10. Briefly analyze the quality of the slide using the phase-contrast microscope at 100× or 200× magnification. Chromosomes on the slide demonstrate different levels of condensation depending on stage of the mitosis: prophase, prometaphase, metaphase, and anaphase. Preparations with >50 chromosome spreads can be considered suitable for FISH.
11. Dip and hold the slide in liquid nitrogen until it stops bubbling. Remove the cover slip from the slide using a razor blade, and immediately transfer the slide to a container of 50% ethanol chilled at −20 °C. Store at 4 °C for at least 2 h for the best dehydration result. If necessary, slides can be stored at this step from several minutes to several days.
12. Dehydrate slides in a series of ethanol (70%, 80%, 100%) at 4 °C for 5 min each, and air-dry at RT. Store dry slides at −20 °C before utilizing them for FISH (*see* **Note 3**).

3.2 Probe Preparation (see Note 6)

3.2.1 Random Primer DNA Labeling

1. If using PCR products as probe, purify the PCR product from the PCR reaction using a QIAquick® PCR purification Kit. Similar kits that remove excess nucleotides can also be used. When using a kit, dissolve the DNA in double-distilled water instead of the Elution buffer suggested in the final step.
2. Add 100 ng template DNA into double-distilled water to a final volume of 13.5 μL in a microcentrifuge tube.
3. Denature the DNA by heating in a boiling water bath for 10 min at 95 °C and chilling quickly in an ice bath.
4. Add the following to the freshly denatured probes on ice:

   1 μL of 1.0 mM dGTP.

   1 μL of 1.0 mM dCTP.

1 μL 1.0 mM dATP.

2 μL of Reaction Mixture from Random Primed DNA Labeling kit.

1 μL of Klenow enzyme.

0.5 μL of 1.0 mM Cy3 or Cy5-dUTP.

Mix and centrifuge briefly.

5. Incubate for 1–20 h (overnight) at 37 °C.
6. Terminate the reaction by adding 1 μL of 0.5 M EDTA to the reaction. Add 1/10 volume of 3 M sodium acetate and 2.5 volume of 100% ethanol and mix by inverting the tubes. Keep at −80 °C or −20 °C for at least 1 h or until probes are needed for hybridization (*see* **Note 7**). If necessary, probes can be left in the freezer for long-term storage.

3.2.2 Nick-Translation DNA Labeling (see **Note 8**)

1. Prepare the following reaction mixture on ice:

5 μL of 10× buffer for DNA Polymerase I.

5 μL of 10× dNTP mixture: 0.5 mM dATP, dCTP, dGTP, and 0.15 dTTP.

4 μL of DNase I freshly diluted to 0.02 units/μL.

1 μL of DNA Polymerase I.

1 μg of Template DNA.

1 μL of Cy3- or Cy5-dUTP.

5 μL of BSA diluted to 0.5 mg/mL.

Add water to final volume 50 μL.

2. Incubate the mix at 15 °C for 3 h.
3. Run 5–10 μL of reaction mixture on an agarose gel to determine the size of digested fragments. Fragments should be 100–600 bp for best hybridization results. If fragments are still larger than this, incubate at 15 °C for additional time.
4. Terminate the reaction by adding 1 μL of 0.5 M EDTA to the reaction. Keep at −20 °C until the hybridization (*see* **Note 7**). If necessary, probes can be left in the freezer for long-term storage.

3.2.3 Extraction of Repetitive DNA Fractions (see **Note 9**)

1. Extract 400–500 mg of the genomic DNA from entire adult mosquitos using a Qiagen Blood and Cell Culture Maxi Kit, and prepare a solution of 100–1000 ng/μL DNA in 1.2× SSC.
2. Denature DNA by placing a safe-lock tube with genomic DNA into a heating block prewarmed to 120 °C for 2 min. High temperature helps to break DNA into 200–500 bp fragments.

Table 1
DNA concentration and reannealing times for preparation of C_0t 2 and C_0t 3 fractions

	DNA concentration (μg/μL)	Reannealing time (min)
C_0t 2	0.1	100
	0.3	33
	0.5	20
	0.7	14
	0.9	11
	1	10
C_0t 3	0.1	150
	0.3	50
	0.5	30
	0.7	21
	0.9	17
	1	15

3. Depending on the DNA concentration, reassociate DNA by placing the tube at 60 °C for 15–150 min to obtain C_0t DNA fractions up to C_0t 3 (Table 1).
4. Place the tube with DNA on ice for 2 min.
5. Transfer the DNA to 42 °C, add preheated 10× S1 nuclease buffer and S1 nuclease to a final concentration of 100 Us per 1 mg of DNA, and incubate for 1 h.
6. Precipitate DNA by adding 0.1 volume of 3 M sodium acetate and 1 volume of isopropanol at RT.
7. Centrifuge at 20,000 × *g* (14,000 rpm) for 20 min at 4 °C.
8. Wash DNA in 70% ethanol, and centrifuge again at 20,000 × *g* (14,000 rpm) for 10 min at 4 °C. Air-dry and dissolve DNA pellet in Tris-EDTA (ethylene diaminetetraacetic acid) buffer.
9. Measure the DNA concentration, and visualize by gel electrophoresis. Usually, the final quantity of repetitive DNA fractions for *Ae. aegypti* represents 35–50% of the original DNA amount.

3.3 FISH

3.3.1 Standard FISH Protocol for Polytene Chromosomes

1. Perform chromosome fixation if slides are more than 2-month-old. If slides are less than 2-month-old, go directly to **step 2**. Fix slides in Carnoy's solution (ethanol/acetic acid in 3:1 ratio) at RT for 10 min and air-dry. Dehydrate slides in 100% ethanol for 10 min and air-dry again.
2. Immerse slides in 1× PBS for 20 min at RT.
3. Incubate slides at RT in 4% formalin (*see* **Note 10**) for 1 min. After incubation in formalin, slide should be immediately immersed in 50% ethanol. Drying the slide prior to dehydration in 100% ethanol damages the chromosome preparation.

4. Dehydrate the slides through an ethanol series of 70%, 90%, and 100% for 5 min each at RT. Air-dry the slides.
5. Centrifuge the tubes of labeled probes at 20,000 × *g* (14,000 rpm) for 10 min at 4 °C. Carefully remove the supernatant, and vacumfuge the tubes for 20 min to dry pellets.
6. Dissolve dry probes in hybridization buffer (60% formamide, 10% dextran sulfate, 1.2× SSC) prewarmed to 37 °C. The amount of hybridization buffer used to dissolve depends on the total amount of DNA probe you are dissolving. Dissolve 250 ng to 1 μg of DNA probe in 20–40 μL of warmed hybridization buffer.
7. In a clean microcentrifuge tube, combine at least 250 ng each of one Cy5-labeled and one Cy3-labeled probes. In situ hybridization on polytene chromosomes is efficient if at least 500 ng of DNA is hybridized on the slide. Vortex and centrifuge the tube of combined probes briefly.
8. Transfer the above prepared solution of combined probes to a chromosome preparation slide and cover with a 22 × 22 mm coverslip. Remove any large air bubbles with gentle pressure.
9. Seal edges of cover slip with rubber cement. Denature the target and probe DNA by placing the slides on the Thermobrite machine at 75–90 °C for 5–10 min. Thermobrite machine does not need to be humid.
10. Transfer the slides to prewarmed humid chambers with 4 × SSC or water in the bottom of the chambers, and incubate at 39 °C for interspecies (e.g., an *An. gambiae* probe with *An. stephensi* chromosomes) or 42 °C for intraspecies hybridization. Allow the probes to hybridize for 3–18 h (usually overnight) (*see* **Note 11**).
11. Carefully remove rubber cement with forceps and coverslip. In a slide jar covered with aluminum foil, wash the slides with 1× SSC at 39 °C after interspecies or 0.2× SSC at 42 °C after intraspecies hybridization for 20 min in 50 mL without agitation.
12. Wash the slides with 1× SSC after interspecies or 0.2× SSC after intraspecies hybridization at RT for 20 min in 50 mL without agitation.
13. Dilute fluorescent dye YOYO-1 100 times in 1× PBS to make a stock solution. Mix 10 μL of 100× diluted YOYO-1 with 90 μL 1× PBS for each slide that you want to stain. The working solution of YOYO-1 is 1000× diluted relative to original concentration.
14. Rinse slide in 1× PBS, and add 100 μL of YOYO-1 in PBS on each slide. Cover with parafilm. Leave at RT for 10 min inside of a slide box or somewhere dark.

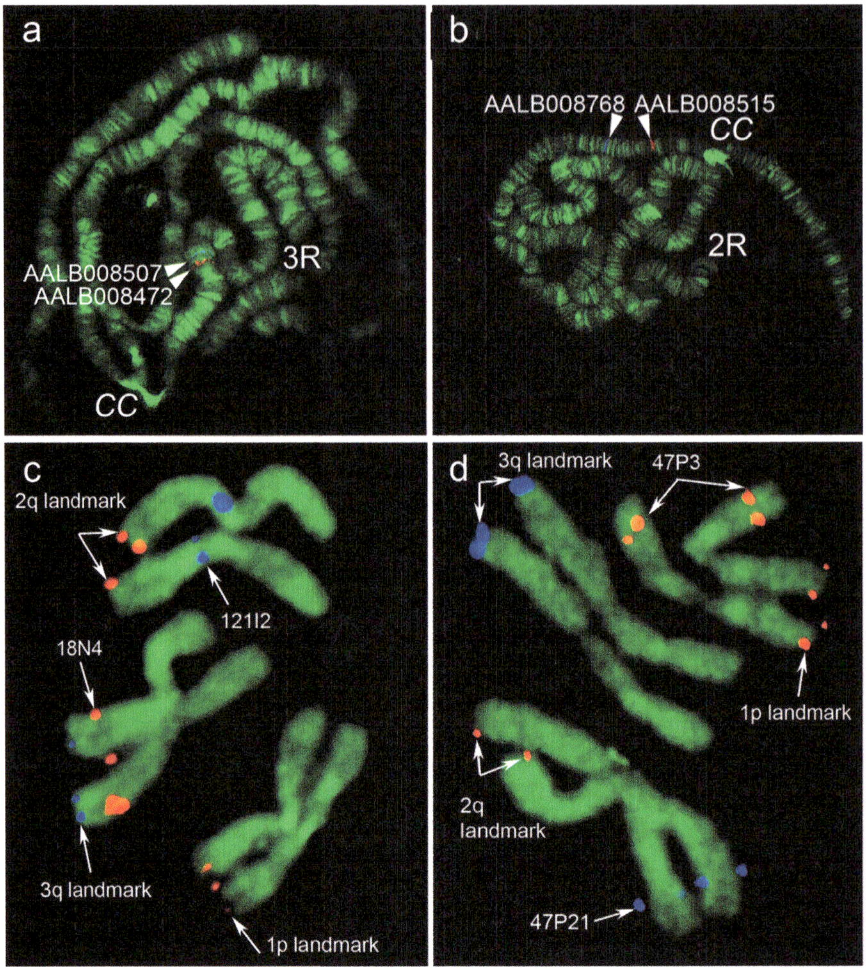

Fig. 4 Examples of FISH on *An. albimanus* polytene [9] (**a**, **b**) and *Ae. aegypti* mitotic (**c**, **d**) chromosomes [13]. Arrows indicate positions of signals on chromosomes including signals of interest and chromosomal landmarks. Chromocenter of polytene chromosomes is indicated as *CC*. 2R, 3R, 1p, 2q, and 3q represent chromosomal arms

15. Rinse in 1× PBS and add 10 μL Prolong Gold antifade reagent, place coverslip on slide, and blot out bubbles. Keep in the slide box at 4 °C. Detect the signals using a confocal or fluorescence microscope, and map them to the polytene chromosomes of *Anopheles* mosquitoes (Fig. 4a, b) (*see* **Note 12**).

3.3.2 FISH Using BAC DNA Clones on Mitotic Chromosomes

1. Incubate slides in 2× SSC for 30 min at 37 °C. Remove residual 2× SSC solution by placing a slide's edge to a paper towel. Dehydrate slides in series of 70%, 80%, and 100% ethanol for 5 min each at RT, and air-dry.
2. Incubate chromosome preparations in 0.1 mg/mL RNase solution under parafilm for 30 min at 37 °C. Wash twice in 2× SSC for 5 min each at 37 °C.

3. Put slides in a jar with 0.01% pepsin and 0.037% HCl solution, and incubate for 5 min at 37 °C. Wash slides in 1× PBS for 5 min at RT.
4. Fix chromosome preparation in a jar with 1% formalin in 1× PBS prepared from 10% neutral-buffered formalin for 10 min at RT. Wash slides in 1× PBS for 5 min at RT. Dehydrate slides in series of 70%, 80%, and 100% ethanol for 5 min each at RT, and air-dry preparations at 37 °C.
5. Denature slides in a jar with prewarmed 70% formamide for 2 min at 72 °C. Dehydrate slides in series of cold (−20 °C) 70%, 80%, and 100% ethanol for 5 min each, and air-dry at 37 °C.
6. Prepare hybridization mixture: 5 μL of labeled BAC DNA, 10 μL of C_0t DNA fractions with final concentration of 0.5 ng/μL, and 5 μL of 1 μg/μL sonicated salmon sperm DNA.
7. Precipitate DNA by adding 0.1 volume of 3 M sodium acetate and 2.5 volumes of ethanol. Keep at −20 °C for 1–3 h.
8. Centrifuge at 20,000 × *g* (14,000 rpm) at 4 °C for 20 min, remove the ethanol, and air-dry the pellet at RT. Thoroughly dissolve the pellet in 10 μL of hybridization buffer: 60% formamide, 20% dextran sulfate, 2× SSC. Denature hybridization mixture for 5 min at 75 °C, and immediately put on ice for 1 min.
9. Prehybridize mixture at 37 °C for 30 min to prevent unspecific hybridization of repetitive DNA to the chromosomes.
10. Place 10 μL of the hybridization mixture on the slide, and cover with a 22 × 22 cover slip. Any air bubbles should be removed with gentle pressure to the coverslip using the tip of a pair of forceps. Glue cover slip around the perimeter using rubber cement. If using multiple probes, add 3–5 μL of each probe in a PCR tube to total volume of 10 μL and mix gently before applying the coverslip.
11. Perform overnight hybridization in a humid chamber at 37 °C (*see* **Note 11**).
12. Remove rubber cement and coverslip from the slide. Wash slide for 2 min in prewarmed Solution 1 (0.4× SSC, 0.3% Nonidet-P40) at 73 °C. Wash slides in Solution 2 (2× SSC, 0.1% Nonidet-P40) for 5 min at RT.
13. Counterstain slide using 0.001 mM YOYO-1 in 1× PBS for 10 min in humid chamber at RT. Mount in a small amount of Prolong Gold antifade reagent with a cover slip.
14. Analyze preparations under a fluorescent microscope using appropriate filter sets at 1000× magnification (Fig. 4c, d) (*see* **Note 13**).

4 Notes

1. To prevent tissues of salivary glands, ovaries, or imaginal discs from drying, add drops of Carnoy's solution (ethanol/acetic acid in 3:1 ratio) while dissecting prior placing them to a drop of 50% propionic acid.
2. **Step 6** in Subheading 3.1.1 is optional. Slides can dry out if left at 4 °C for extended periods of time. Leaving them at 4 °C for longer than overnight is not recommended.
3. Slides that are kept protected from dust and debris can be used for FISH at least within a year after the preparations are made.
4. Using modified Carnoy's solution (methanol/acetic acid, 3/1) for the fixation of ovaries provides a better extraction of fat from the follicles. A standard Carnoy's solution (ethanol/acetic acid in 3:1 ratio) can also be used instead.
5. **Step 14** in Subheading 3.1.2 is optional. Slides can be stored at this step from several minutes to several days, if necessary.
6. Different DNA probes can be used for physical mapping of the mosquito genomes, such as BAC clones, cDNA, PCR amplified probes, and others. Random Primer labeling is used for fragments shorter than 1 Kb. Nick-translation labeling is used for fragments longer than 1 Kb (1–150 Kbs).
7. Fluorescently labeled probes should be protected from light. Even where it is not explicitly stated, make efforts to protect probes from light.
8. This protocol can be scaled down by ½ to accommodate 500 ng of template DNA. Final concentrations of DNase I and DNA Polymerase I have to be optimized based on factors including: initial size of template DNA and reaction time. Larger template size generally requires more DNAase I. The reaction time influences the final fragment size because DNA Polymerase I stars losing its activity earlier than DNAse I.
9. Performing FISH of a BAC clone DNA probe on chromosomes from Culicinae mosquitoes such as *Ae. aegypti* and *Cx. quinquefasciatus* requires using unlabeled repetitive DNA fractions to block unspecific hybridization of the DNA repeats to the chromosomes. The reassociation of single-strand DNA fragmented into pieces of several hundred bp follows a C_0t curve where C_0 is the initial concentration of single-stranded DNA and t is the reannealing time. DNA fractions with C_0t values equal to 10^{-4}–10^{-1} or 10^{0}–10^{2} are considered as highly and moderately repetitive, respectively.

10. Prepare 4% formalin fresh: mix 5 mL of 10× PBS, 20 mL of 10% neutral-buffered formalin and 25 mL of water. Formalin is hazardous and should be handled carefully. Avoid breathing gas or dust during preparation: wear gloves and other personal protective equipment when handling. Formalin solution should not be dumped down drains.
11. Because there are fluorescently labeled probes on the slide(s), humid chambers should be impermeable to light. It is advisable to cover the humid chamber with aluminum foil to minimize light impact on the probes.
12. For positioning of DNA probes on cytogenetic maps use chromosomal landmarks and banding patterns as indicated in cytogenetic descriptions of the maps.
13. Most chromosomes in mosquitoes are metacentric meaning that they have almost equal lengths of the chromosomal arms. For the correct assignment of DNA probes to the chromosomal arms, additional molecular landmarks must be used. BAC clones or other probes with known telomeric positions in the chromosomes can be used as molecular landmarks helping to identify the correct chromosomal arm position of a probe of interest (Fig. 4c, d). For the mapping first consider large negative bands to identify the position of a probe within a chromosomal region. When a chromosomal region is identified, assign the signal to a chromosomal band.

Acknowledgment

Physical mapping of *Anopheles* was supported by Russian Science Foundation 15-14-20011 to Igor V. Sharakhov and physical mapping of *Aedes* and *Culex* was supported the National Institutes of Health grants 1R21AI121853 and 1R21AI123967 to Maria V. Sharakhova. We thank Phillip George for providing images of ovary dissection and Melissa Wade for proofreading the text.

References

1. Tolle MA (2009) Mosquito-borne diseases. Curr Probl Pediatr Adolesc Health Care 39:97–140
2. Sharakhov IV, Sharakhova MV (2008) Cytogenetic and physical mapping of mosquito genomes. In: Verrity JF, Abbington LE (eds) Chromosome mapping research developments. Nova Science Publishers, Inc, New York, pp 35–76
3. George P, Sharakhova MV, Sharakhov IV (2010) High-resolution cytogenetic map for the African malaria vector Anopheles gambiae. Insect Mol Biol 19:675–682
4. Jiang X, Peery A, Hall AB, Sharma A, Chen XG, Waterhouse RM et al (2014) Genome analysis of a major urban malaria vector mosquito, *Anopheles stephensi*. Genome Biol 15:459
5. Sharakhova MV, Xia A, Tu Z, Shouche YS, Unger MF, Sharakhov IV (2010) A physical map for an Asian malaria mosquito, *Anopheles stephensi*. Am J Trop Med Hyg 83:1023–1027

6. Wei Y, Cheng B, Zhu G, Shen D, Liang J, Wang C et al (2017) Comparative physical genome mapping of malaria vectors *Anopheles sinensis* and *Anopheles gambiae*. Malar J 16:235
7. Artemov GN, Sharakhov MV, Naumenko AN, Karagodin DA, Baricheva EM, Stegniy VN et al (2015) A standard photomap of ovarian nurse cell chromosomes in the European malaria vector *Anopheles atroparvus*. Med Vet Entomol 29:230–237
8. Neafsey DE, Waterhouse RM, Abai MR, Aganezov SS, Alekseyev MA, Allen JE et al (2015) Mosquito genomics. Highly evolvable malaria vectors: the genomes of 16 *Anopheles* mosquitoes. Science 347:1258522
9. Artemov GN, Peery AN, Jiang X, Tu Z, Stegniy VN, Sharakhova MV, Sharakhov IV (2016) The physical genome mapping of *Anopheles albimanus* corrected scaffold misassemblies and identified inter-arm rearrangements in genus Anopheles. G3 (Bethesda)
10. Timoshevskiy VA, Severson DW, Debruyn BS, Black WC, Sharakhov IV, Sharakhova MV (2013) An integrated linkage, chromosome, and genome map for the yellow fever mosquito *Aedes aegypti*. PLoS Negl Trop Dis 7:e2052
11. Sharakhova MV, Timoshevskiy VA, Yang F, Demin S, Severson DW, Sharakhov IV (2011) Imaginal discs: a new source of chromosomes for genome mapping of the yellow fever mosquito *Aedes aegypti*. PLoS Negl Trop Dis 5:e1335
12. Naumenko AN, Timoshevskiy VA, Kinney NA, Kokhanenko AA, deBruyn BS, Lovin DD et al (2015) Mitotic-chromosome-based physical mapping of the culex quinquefasciatus genome. PLoS One 10:e0115737
13. Timoshevskiy VA, Kinney NA, deBruyn BS, Mao C, Tu Z, Severso DW (2014) Genomic composition and evolution of *Aedes aegypti* chromosomes revealed by the analysis of physically mapped supercontigs. BMC Biol 12:27
14. Xia A, Peery A, Kamali M, Liang J, Sharakhova MV, Sharakhov IV (2013) Fluorescence in situ hybridization to the polytene chromosomes of anopheles mosquitoes. Bio-protocol 3:e860
15. Timoshevskiy VA, Sharma A, Sharakhov IV, Sharakhova MV (2012) Fluorescent in situ hybridization on mitotic chromosomes of mosquitoes. J Vis Exp (67):4215
16. Sharakhova MV, George P, Timoshevskiy V, Sharma A, Peery A, Sharakhov IV (2015) Mosquitoes (Diptera). In: Sharakhov IV (ed.), Protocols for cytogenetic mapping of Arthropod genomes. CRC Press, Taylor & Francis Group, Boca Raton, London, New York, pp 93–170
17. Clements AN (1992) The biology of mosquitoes: development, nutrition and reproduction. Chapman & Hall, London

Chapter 14

Target-Enriched Endosymbiont Sequencing (TEEseq): A New High-Throughput Sequencing Approach Applied to the Comprehensive Characterization of Endosymbionts

Hannes Schuler, Jacqueline A. Lopez, Meredith M. Doellman, Glen R. Hood, Scott P. Egan, and Jeffrey L. Feder

Abstract

Intracellular bacteria are ubiquitous in the insect world, with perhaps the best-studied example being the alphaproteobacterium, *Wolbachia*. Like most endosymbionts, *Wolbachia* cannot be cultivated outside of its host cells, hindering traditional microbial characterization techniques. Furthermore, multiple *Wolbachia* strains can be present within a single host, and certain strains can be present in densities below the detection limit of current methods. To date, *Wolbachia* has most commonly been studied using polymerase chain reaction (PCR) amplification and Sanger DNA sequencing by targeting specific genes in the bacterium's genome. PCR amplification and Sanger sequencing of multiple *Wolbachia* strains requires analysis of individually cloned sequences, which is resource and labor intensive. To help mitigate these difficulties, we present a modified double digest restriction site associated DNA sequencing (ddRADseq) approach to target and sequence in parallel multiple genes by adding restriction enzyme recognition sites to gene-specific PCR primers. Adopting this strategy allows us to uniquely tag and sequence amplicons from multiple hosts simultaneously on an Illumina MiSeq platform. Our approach represents an efficient and cost-effective method to characterize multiple target genes in population surveys.

Key words Wolbachia, Endosymbiont, Microbial community, Metabarcoding, PCR enrichment, ddRAD, Illumina sequencing

1 Introduction

Maternally inherited endosymbionts are widespread among insects [1]. A key factor responsible for their prevalence is their ability to alter the reproduction of their hosts in ways favoring their own spread [2]. The alphaproteobacterium *Wolbachia* is perhaps the most successful of all endosymbionts, being present in about 50% of all arthropod species [3]. Genetic and molecular characterization

Electronic supplementary material: The online version of this chapter (https://doi.org/10.1007/978-1-4939-8775-7_14) contains supplementary material, which is available to authorized users.

Susan J. Brown and Michael E. Pfrender (eds.), *Insect Genomics: Methods and Protocols*, Methods in Molecular Biology, vol. 1858, https://doi.org/10.1007/978-1-4939-8775-7_14, © Springer Science+Business Media, LLC, part of Springer Nature 2019

of most endosymbionts, including *Wolbachia*, is not trivial, however, as the vast majority of these bacteria cannot be cultured in vitro. Consequently, whole genome sequencing has been performed on just a few *Wolbachia* strains to date, e.g., [4–6]. In addition, our knowledge of the population genetics of *Wolbachia* has been largely based on PCR amplification and Sanger DNA sequencing of six target genes, specifically the *Wolbachia* surface protein (*wsp*) [7] and five other genes used for the multilocus sequence typing (MLST) approach [8]. The MLST system uses five conserved genes across the *Wolbachia* genome that are amplified, Sanger sequenced, and used for phylogenetic characterization of different strains. Today MLST is the most commonly used tool to genotype this endosymbiont [9–13].

The PCR and Sanger DNA sequencing approach has significant limitations. Since multiple *Wolbachia* strains can be present within a single host, e.g., [14], it can be difficult to characterize less common strains unless cloning and extensive sequencing are performed. Indeed, rare strains are often discarded as sequencing error and overlooked by conventional Sanger sequencing of *Wolbachia*-specific genes. This is a significant problem, as it is known that certain *Wolbachia* strains are present in low abundance in their hosts [15] and multiple sequence profiles can be resolved only when individual amplicons are cloned, sequenced, and analyzed. Moreover comparative genomics of different *Wolbachia* genomes [16] showed that the MLST genes are too conservative and not suitable for discrimination of closely related strains [17]. Therefore more than 250 additional single copy genes have been proposed for a fine scale strain differentiation [16, 17]. Finally, characterization of multiple genes in multiple populations by Sanger sequencing can be time consuming and relatively costly, particularly when it requires the cloning of individual amplicons.

Here, we present the Target-enriched endosymbiont sequencing (TEEseq) approach, a modification of the double digest restriction site associated DNA sequencing (ddRADseq) approach commonly utilized in population genetic studies [18–20] as a means to enhance the efficiency of genetic screening of endosymbionts, in general, and of *Wolbachia*, in particular (Fig. 1). TEEseq involves first PCR amplifying DNA isolated from a given host sample using modified *Wolbachia* gene-specific primers. At its 5′ end, the forward primer has been designed to include an EcoR1 restriction enzyme recognition site, while the reverse primer incorporates a Not1 restriction enzyme recognition site (Fig. 1a). Next, the amplicons (Fig. 1b) are digested by EcoR1 and Not1 restriction enzymes. Then, barcoded adapters containing Illumina sequencing primer sites are ligated to the sticky ends of the digested PCR DNA fragments (Fig. 1c). The barcodes distinguish amplicons generated from different host samples. After ligation, a second PCR enriches for the adapter-ligated amplicons. Lastly, amplicons generated from

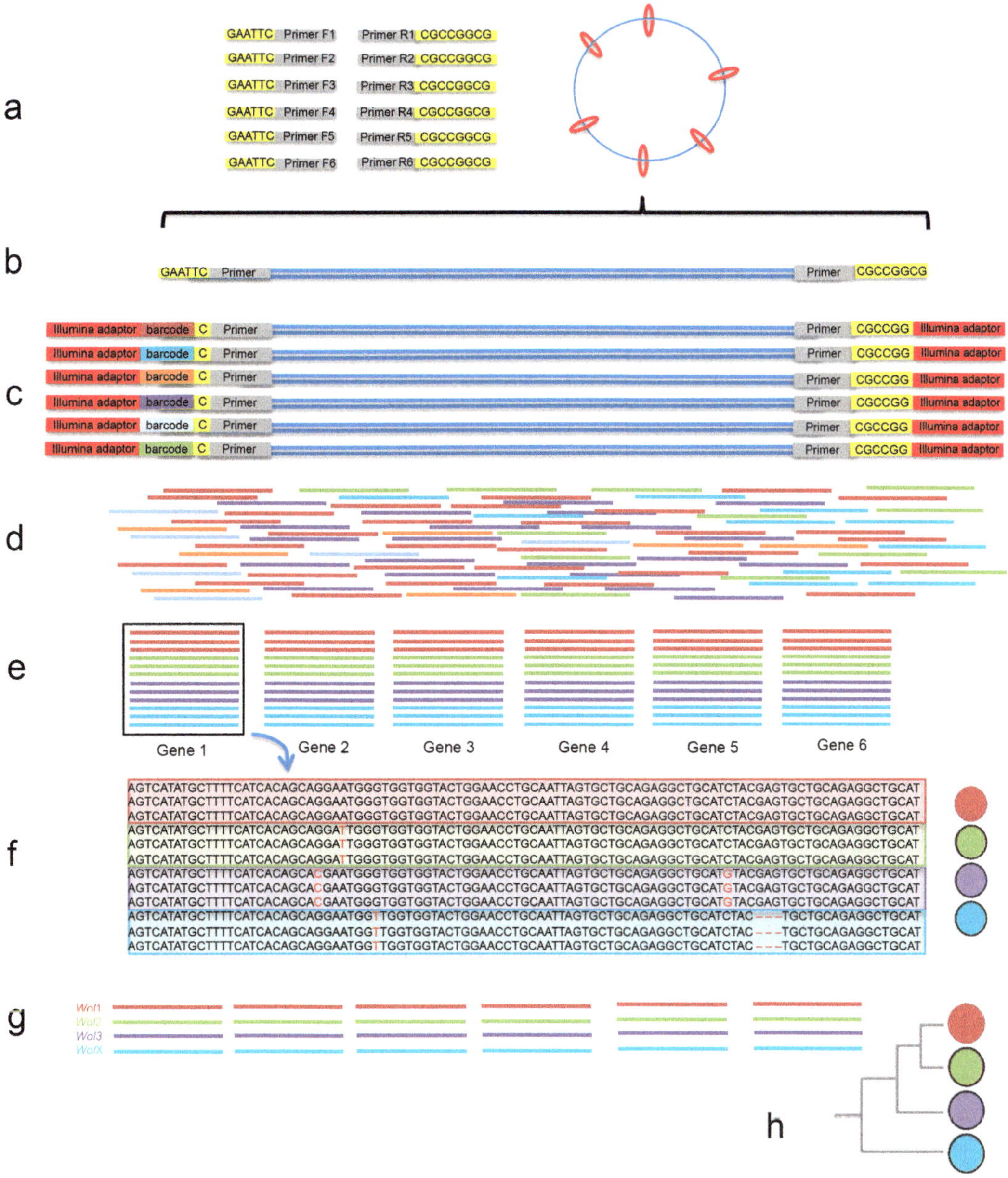

Fig. 1 Schematic overview of the TEEseq approach: (**a**) Prior to PCR, forward and reverse primers are designed to amplify a 300 bp region of the *wsp* gene and the five MLST genes of *Wolbachia* (positions in the circular genome of the bacterium schematically depicted by red ovals). These primers include EcoR1 and Not1 restriction cut sites on their ends, respectively. (**b**) PCR is performed for all six of the target genes separately using total DNA isolated from an individual host sample, with multiple reactions from different host specimens done in parallel. (**c**) Each PCR amplicon is separately digested with restriction enzymes EcoR1 and Not1, generating an overhang at each end of the fragment, after which the fragments are ligated to adapters containing specific 8–10 bp barcodes (different colors—allowing for pooling of >96 host samples) and Illumina sequencing primers (red bars—for PCR enrichment of adapter-ligated fragments) (**d**) and sequence with the Illumina MiSeq System. (**e**) Following sequencing, reads are stacked by matching loci, (**f**) aligned and assigned to a given host according to the barcode, and scored, allowing for (**g**) the frequencies of variants at the target loci to be determined for population analysis, and (**h**) the prevalence and distributions of strains to be assessed and their phylogenetic relationships to be analyzed

different target genes and host samples are pooled, purified, and sequenced on an Illumina MiSeq platform (Fig. 1d). Compared to the PCR and Sanger DNA sequencing approach, the TEEseq approach costs less per sample, generates data faster, and provides potentially up to 30 million paired-end sequencing reads per MiSeq lane. The large number of reads provides a high coverage per gene and therefore ultra deep sequencing of specific genes allow the detection of rare sequence variants in mixed populations.

Using our novel approach, we sequenced all six genes from 190 individual flies of the genus *Rhagoletis*. The initial ~8 million reads were demultiplexed, quality filtered, trimmed, and mapped into different genes using *mothur* [21]. After quality filtering we obtained between 474,000 and 750,000 reads per gene that resulted in a minimum of 2000 reads per *Wolbachia* gene per individual (Schuler et al. unpubl. data). Thus our pilot study highlights the potential of TEEseq for the parallel high-resolution characterization of different *Wolbachia* genes. We note that the high coverage allows for the inclusion of more than just the proposed six genes documented here. In summary, TEEseq is a highly flexible approach that can be expanded to a variety of target genes from other systems and thus provides a new cost-efficient tool for the high-resolution characterization of different genes.

2 Materials

2.1 General Materials and Equipment

All procedures are performed in a 96-well plate format. Single channel pipettors are used to prepare master mixes; multichannel pipettors are used when working with 96-well plates to reduce time and labor. Certified low retention, nuclease-free, and PCR-free 96-well plates, centrifuge tubes, and aerosol-barrier filter tips are used for all steps. All reagents and chemicals are molecular-grade. Oligonucleotide primers are purchased from any oligonucleotide synthesis company; choose standard desalting purification and the smallest allowable scale of production. Equipment and materials are:

1. Two boxes of iProof High-Fidelity DNA Polymerase, 500 U (2 U/μL). Each box includes 250 μL iProof Polymerase, 4 × 1.5 mL 5× iProof HF Buffer, 2 × 1.5 mL 50 mM magnesium chloride ($MgCl_2$), and 500 μL DMSO solution; store at −20 °C (*see* **Note 1**).
2. 150 μL of dNTP mix (2.5 mM of each dATP, dCTP, dGTP, and dTTP) prepared from 100 mM dNTP set; store at −20 °C.
3. 1 mL of dNTP mix (10 mM of each dATP, dCTP, dGTP, and dTTP) prepared from 100 mM dNTP set; store at −20 °C.

4. 30 mL of 0.1× TE Solution (1mM Tris–HCl, 0.1mM EDTA) prepared from 1 M Tris–HCl, pH 9.0 and 0.5 M EDTA Solution, pH 8.0; store at room temperature.
5. 100 mL of nuclease-free water; store at room temperature.
6. 150 μL of 1 mg/mL bovine serum albumin (BSA). Dilute 10 mg/mL bovine serum albumin with nuclease-free water. Store at −20 °C.
7. 500 μL of 1 M sodium chloride (NaCl). Dilute 5 M sodium chloride with nuclease-free water; store at room temperature.
8. Minimum of 32 boxes of extra-long P10, 26 boxes P200, and 1 box P1000 pipette tips.
9. Minimum of 26 × 96-well PCR plates.
10. Minimum of 30 × PCR adhesive seals.
11. At least six 0.2 mL PCR tubes, six 8-well PCR strips with strip caps, twenty-five 1.5 mL microcentrifuge tubes, five 2.0 mL microcentrifuge tube, twenty 5 mL centrifuge tubes, and twenty-four 50 mL centrifuge tubes.
12. Minimum of five 25 mL reagent reservoirs.
13. Pipettors: Single channel set of P10, P200, and P1000 and 8-channel P10 and P100.
14. Minimum of two 96-well PCR thermal cyclers with heated lid are required for this protocol.
15. Centrifuges and vortex: Benchtop plate centrifuge, microcentrifuge for 1.5 mL microcentrifuge tubes, minicentrifuge for PCR 8-tube strips, and mini vortex.

2.2 *Wolbachia* Gene-Specific PCR

1. Total DNA isolated from host samples are arrayed in a 96-well plate containing 95 unique DNA samples and one no template control (i.e., nuclease-free water); each sample well contains 15 μL of DNA from a single individual. Add 15 μL of nuclease-free water to the no-template control well. Store at −20 °C (*see* **Note 2**).
2. The *Wolbachia* gene-specific PCR uses six different primer pairs (Table 1) to amplify a fragment of the *wsp* gene and five MLST genes [7, 8]. EcoR1 restriction recognition site is added to the 5′ end of the forward primer. Not1 restriction recognition site is added to the 5′ end of the reverse primer (*see* **Note 3**). All primary primer stocks (100 μM) are diluted with nuclease-free water to 10 μM working stocks. A minimum volume of 50 μL of each working primer stock is required. Store at −20 °C.

Table 1
Modified primers targeting a ~300 bp internal fragment of the *wsp* gene and five MLST genes

Gene	Name	Sequence 5′–3′	Fragment length
wsp	wsp-F	TA**GAATTC**TTGGTTAYAAAATGGAYGAYAT	310
	wsp-R	TA**GCGGCCGC**GAGCTCCAGCAWARAGTTTG	
coxA	coxA-F	TA**GAATTC**GTTRATTGTTGCCTTACCRG	330
	coxA-R	TA**GCGGCCGC**ACRCCMGTTATAACACCRAT	
ftsZ	ftsZ-F	TA**GAATTC**CAYATGCTYTTYATCACAGCA	315
	ftsZ-R	TA**GCGGCCGC**TGAYCATCARRTCAGTTACTC	
fbpA	fbpA-F	TA**GAATTC**GCWTRGCTGTYGGRTTTACT	321
	fbpA-R	TA**GCGGCCGC**AYYACTATTCTTTTYCCTGMAA	
hcpA	hcpA-F	TA**GAATTC**GCTGGAGAAARTTAYGARGA	293
	hcpA-R	TA**GCGGCCGC**CRTCRCGYACTYTACCAAARTC	
gatB	gatB-F	TA**GAATTC**GTTTCAGARCCAGATCTYCG	307
	gatB-R	TA**GCGGCCGC**TARTCRCTTGMATCYTCTTT	

The EcoR1 restriction site in the forward primer and the Not1 restriction site in the reverse primer are shown in bold text

Table 2
Illumina primer sequences

	Name	Sequence 5′–3′
ILLPCR	ILLPCR-1	AATGATACGGCGACCACCGAGATCTACACTCTTTCCCTACACGACG CTCTTCCGATCT
	ILLPCR-2	CAAGCAGAAGACGGCATACGAGATCGTGATGTGACTG

2.3 Restriction Digest with EcoR1 and Not1, and Ligation of Illumina-Compatible Adapters

1. 80 μL EcoR1, 10,000 U (20,000 units/mL); store at −20 °C.
2. 80 μL Not1, 500 U (10,000 units/mL); store at −20 °C.
3. 130 μL T4 DNA Ligase, 20,000 U (400,000 U/mL); store at −20 °C.
4. 825 μL 10× T4 DNA Ligase Buffer. Store at −20 °C (*see* **Note 4**).

2.4 PCR Enrichment of Adapter-Ligated Library

1. The Illumina PCR primers (Table 2) amplify DNA fragments with ligated EcoR1 and Not1 adapter sequences (*see* **Note 5**). In two 1.5 mL microcentrifuge tubes, mix 60.0 μL of ILLPCR-1 primer (100 μM) and 60.0 μL of ILLPCR-2 primer (100 μM) with 1080.0 μL of nuclease-free water (*see* **Note 6**). Store the ILLPCR primer mix (5 μM each primer) at −20 °C.

2.5 Purification of Enriched Library

1. Agencourt AMPure XP, 5 mL (Beckmann Coulter). At least 5 mL is required. Equilibrate the bottle of Agencourt AMpure XP to room temperature for 30 min. Immediately prior to use,

vortex the bottle for 2 min to distribute the magnetic particles in solution.

2. 25 mL 70% ethanol is prepared immediately prior to use from Ethanol 200 Proof Absolute and nuclease-free water. Take care to avoid ethanol evaporation; lower ethanol content washes away DNA from the magnetic beads. Discard and freshly prepare 70% ethanol before the experiment.
3. Minimum of one DynaMag-2 Magnet (Invitrogen Life Technologies) is required.

2.6 Validation of Library

1. Access to a capillary electrophoresis system. We validated the fragment length of our library using Agilent DNA 7500 Kit on Agilent Bioanalyzer 2100 System.
2. Access to a quantitative PCR system. We quantitated our completed library using KAPA Library Quantitation Kit for Illumina libraries on BioRAD CFX Real-Time Detection System.

2.7 Illumina Sequencing

Access to a Next Generation Sequencing (NGS) provider. We sequenced our completed library with the Illumina MiSeq System (Illumina Inc.) using MiSeq Reagent Kit v2 500-cycle (Illumina Inc.) and PhiX Control v3 (Illumina Inc.).

3 Methods

From a single DNA sample, target genes are PCR amplified in parallel using gene-specific primers containing EcoRI and NotI restriction enzyme recognition sites on their ends. The PCR products are digested with EcoRI and NotI, ligated to barcoded Illumina-compatible adapters, and amplified with adapter-specific primers. Libraries constructed from different total DNA samples are pooled and sequenced on Illumina MiSeq System.

Our full-scale TEEseq protocol processes 96 DNA samples with six gene-specific primer pairs. Six gene-specific 96-well reaction plates are set up in **steps 3.1** and **3.2** and twelve 96-well PCR enrichment plates in **step 3.3**. We recommend starting with 96 DNA samples and one gene-specific primer pair to construct one gene-specific reaction plate and two PCR enrichment plates in **steps 3.1–3.3**. Then, repeat **steps 3.1–3.3** five more times for the remaining gene-specific primer pairs to make all six target genes for each DNA sample in **step 3.4**.

When the step says "centrifuge briefly," centrifuge the 96-well plate, 8-well PCR strip, or microcentrifuge tube at 500 × *g* for 30 s at room temperature. All master mix volumes are scaled by 1.25 times to allow for pipetting loss. Once the master mix is prepared, mix by slowly pipetting up and down 10 times. Keep enzymes, master mix components and 96-well reaction plates on ice unless

Table 3
Master mix for gene-specific PCR

Component	Amount per rxn (μL)	Amount per 96-well plate (μL)
Nuclease-free water	6	720
5× iProof HF Buffer	2	240
BSA, 1 mg/mL	0.1	12
dNTPs, 2.5 mM each	0.2	24
Gene-specific forward primer, 10 μM	0.3	36
Gene-specific reverse primer, 10 μM	0.3	36
iProof HF Polymerase, 2 U/μL	0.1	12
Final volume	9	1080

otherwise noted. Thaw master mix components (buffers, dNTP mix, and primers) on ice unless otherwise noted. Use an 8-well PCR strip with strip cap and 8-channel pipettor to dispense master mix into each well of the 12 columns of the 96-well PCR plate.

3.1 PCR Enrichment of *Wolbachia* Fragments

The *Wolbachia* PCR primer pairs each amplify a different gene fragment of *Wolbachia* [7]. A separate reaction is prepared for each primer pair per host DNA sample (*see* **Note 7**).

1. Remove the DNA Plate, 5× iProof HF Buffer, dNTP mix, and BSA from storage, centrifuge briefly, and place on ice.
2. On ice, prepare the gene-specific PCR master mix in a microcentrifuge tube (1.5 mL) according to Table 3.
3. Transfer 115 μL of master mix to each well of an 8-well PCR strip. Centrifuge briefly, and place on ice.
4. Prepare the gene-specific PCR reaction plate. With 8-channel P10 pipettor, add 9 μL gene-specific PCR master mix to the bottom of each well of a 96-well PCR plate, and add 1 μL of DNA from the DNA sample plate. Seal the PCR reaction plate with PCR adhesive seal. Centrifuge briefly and place the plate into the thermocycler.
5. Follow the thermocycler conditions outlined in Table 4 (*see* **Note 8**).
6. When the PCR cycle is complete, remove the plate and centrifuge briefly.
7. Check amplification success by gel electrophoresis (*see* **Note 9**).

Table 4
Thermocycler conditions for gene-specific PCR

Steps	Time/cycles	Block temperature (°C)	Lid temperature (°C)
1. Denaturation	4 min	94	105
2. Denaturation	30 s	94	105
3. Annealing	1 min	55	105
4. Extension	1 min	72	105
5. Repeat steps 2–4	33 cycles more		
6. Final extension	10 min	72	105
7. Hold	Infinity	4	Heated lid off

Table 5
EcoR1 and Not1 adapter sequences

	Name	Sequence 5′–3′
EcoRl adapter	EcoRl-A	CTCTTTCCCTACACGACGCTCTTCCGATCTbarcodeC
	EcoRl-B	TGTGAGAAAGGGATGTGCTGCGAGAAGGCTAGA barcodeGTTAA
Notl adapter	Notl-A	GCAGAAGACGGCATACGAGATCGTGATGTGACTGG AGTTCAGACGTGTGCTCTTCCGATC
	Notl-B	GGCCGATCGGAAGAGCACACGTCTGAACTCCAGTC ACATCACGATCTCGTATGCCGTCTTCTGCTTG

3.2 Preparation of Adapters

3.2.1 EcoR1 Adapter

Barcoded EcoRl adapter: The EcoRl adapter is a double-stranded oligonucleotide consisting of A and B strands that contain an Illumina sequencing primer site, an 8 to 10-bp barcode sequence and restriction site overhang [18] (Table 5; Supplemental Table 1). Mix 1 μL each of EcoRl A and B strands (100 μM) in each well of a 96-well PCR Plate, with 98 μL of water, heat to 95 °C for 5 min, and then slowly cool to room temperature (*see* **Note 10**). Store the 1 μM EcoRl barcode adapter plate at −20 °C.

3.2.2 Not1 Adapter

The Notl adapter is a double-stranded oligonucleotide (Table 5) consisting of A and B strands that contain the Illumina paired-end (PE) sequencing primer site and Notl restriction site overhang. In six 0.2 mL PCR tubes, mix 1.5 μL of each Notl A and B strands (100 μM) with 148.5 μL of nuclease-free water, heat to 95 °C for 5 min, and then slowly cool to room temperature (*see* **Note 11**). Store Notl adapter (1 μM) at −20 °C.

3.3 Restriction Digest and Ligation of Barcodes and Illumina-Compatible Adapters

During this step, EcoR1 and Not1 restriction enzymes digest an aliquot of the gene-specific PCR product, and T4 Ligase attaches the Illumina-compatible adapters. Use the same EcoR1 barcoded adapter for all six gene fragments derived from the same DNA sample.

1. Remove from storage and thaw the 10× T4 Buffer and 1 mg/mL BSA solution at room temperature. Retrieve 1M sodium chloride solution and nuclease-free water from storage. EcoR1 and Not1 restriction enzymes should stay in the freezer until immediately prior to addition to the master mix.
2. Transfer 6 μL of gene-specific PCR product from the PCR reaction plate from **step 3.1** to a new 96-well plate and place on ice. Store the remaining volume of the PCR reaction plate at −20 °C.
3. On ice, prepare the Restriction Digest master mix in a microcentrifuge tube (1.5 mL) according to Table 6.
4. Transfer 39 μL of the Restriction Digest master mix into each well of a PCR 8-strip tube. Centrifuge briefly and place on ice.
5. Using an 8-channel P10 pipette, add 2.6 μL of master mix to each well of the plate containing 6 μL of PCR product. Pipette gently 3 times to rinse master mix from the tip. Seal the Restriction Digest plate with a PCR adhesive seal. Centrifuge and place the plate into the thermocycler.
6. Follow the thermocycler conditions outlined in Table 7.
7. Retrieve the EcoR1 Adapter plate and Not1 Adapter (*see* **Note 12**) from storage and thaw on ice 30 min prior to completion of the restriction digest thermocycler program. Centrifuge the EcoR1 Adapter plate before removing the seal. T4 DNA Ligase should stay in the freezer until immediately prior to addition to the master mix.

Table 6
Master mix for restriction digest

Component	Amount per rxn (μL)	Amount per 96-well plate (μL)
Water	0.78	93.6
10× T4 Buffer	1	120
NaCl, 1 M	0.52	62.4
BSA, 1 mg/mL	0.1	12
EcoR1 enzyme, 20,000 U/mL	0.1	12
Not1 enzyme, 10,000 U/mL	0.1	12
Final volume	2.6	312

Table 7
Thermocycler conditions for restriction digest

Steps	Time	Block temperature (°C)	Lid temperature (°C)
Step 1	2 h	37	105
Step 2	20 min	65	105
Step 3	Infinity	4	Heated lid off

Table 8
Master mix for ligation of Illumina-compatible adapters

Component	Amount per rxn (μL)	Amount per 96-well plate (μL)
Water	0.07	8.64
10× T4 Buffer	0.1	12
NaCl, 1 M	0.05	6
BSA, 1 mg/mL	0.05	6
Notl adapter, 1 μM	1	120
T4 DNA ligase, 400,000 U/mL	0.17	20.1
Final volume	1.44	172.74

8. When the thermocycler program completes, remove the restriction digest plate from the thermocycler and centrifuge briefly.
9. On ice, prepare the Adapter Ligation master mix in a micro centrifuge tube (1.5 mL) according to Table 8.
10. Transfer 21 μL of the Adapter Ligation master mix into each well of an 8-well PCR-strip tube. Centrifuge briefly and place on ice.
11. Remove the PCR adhesive seal from the restriction digest plate. Using an 8-channel P10 pipette, add 1.4 μL of Adapter Ligation master mix to each well of the 96-well plate containing the digested PCR products. Pipette gently 3 times to rinse master mix from the tip.
12. Using an 8-channel P10 pipette, add 1 μL of EcoR1 Adapter to each well of the 96-well plate. Pipette gently 3 times to rinse adapter from the tip. Seal both the EcoR1 Adapter plates and Adapter Ligation plate with a new PCR adhesive seal. Return the EcoR1 Adapter plate to −20 °C storage. Centrifuge the Adapter Ligation plate and place it into the thermocycler.
13. Follow the thermocycler conditions outlined in Table 9.

Table 9
Thermocycler conditions for the adapter ligation

Steps	Time	Block temperature (°C)	Lid temperature (°C)
Step 1	2 h	16	20
Step 2	Infinity	4	Heated lid off

14. When the thermocycler program completes, remove the plate. Centrifuge briefly and place on ice. Remove the PCR adhesive seal. Dilute each reaction with 40 μL of 0.1× TE (pH 7.5–8.0).
15. Seal the plate with a new PCR adhesive seal, centrifuge, and place on ice (*see* **Note 13**).

3.4 PCR Enrichment of Illumina-Compatible Library

In this step, two separate 20 μL PCR enrichment reactions are assembled for each well of the diluted adapter-ligated DNA plate.

1. Remove the 5× iProof HF Buffer, 50mM magnesium chloride, DMSO, ILLPCR primer mix, dNTP mix, and nuclease-free water from storage. iProof High-Fidelity DNA Polymerase should stay in the freezer until immediately prior to addition to the master mix.
2. Prepare two PCR enrichment reaction plates for the diluted adapter-ligated DNA plate. Transferring 4 μL of restriction digested, adapter-ligated DNA into two new 96-well PCR plates.
3. On ice, prepare the PCR enrichment master mix in a centrifuge tube (5 mL) according to Table 10.
4. Set a P1000 to 1000 μL and pipette up and down gently 5 times to mix. Avoid introducing bubbles. Transfer 120 μL of the PCR enrichment master mix into each well of four 8-well PCR-strip tubes. Centrifuge briefly and place on ice.
5. Using an 8-channel P100 pipette, add 16 μL of the PCR enrichment master mix to each well of the two PCR plates containing 4 μL of the diluted adapter-ligated DNA. Pipette gently 3 times to rinse master mix from the tip.
6. Seal each PCR plate with a PCR adhesive seal and centrifuge briefly.
7. Follow the thermocycler conditions outlined in Table 11.
8. When thermocycler program completes, remove plates from the thermocycler and centrifuge briefly.
9. Transfer the products of both PCR enrichment plates into a single centrifuge tube (5 mL). Samples can be pooled by different amplification success of **step 3.1** (*see* **Note 9**). Place on ice and proceed to the next step.

Table 10
Master mix for PCR enrichment of Illumina-compatible library

Component	Amount per rxn (μL)	Amount for two 96-PCR plates (μL)
Water	18.8	2256
5× iProof HF Buffer	8	960
ILLPCR primer mix, 5 μM each	2.6	312
dNTP mix, 10 mM each	0.8	96
$MgCl_2$, 50 mM	0.8	96
DMSO	0.6	72
iProof Polymerase, 2 U/μL	0.4	48
Final volume	32	3840

Table 11
Thermocycler conditions for PCR enrichment of Illumina-compatible library

Steps	Time/cycles	Block temperature (°C)	Lid temperature (°C)
1. Denaturation	30 s	98	105
2. Denaturation	20 s	98	105
3. Annealing	30 s	60	105
4. Extension	40 s	72	105
5. Repeat steps 2–4	29 cycles more		
6. Final extension	10 min	72	105
7. Hold	Infinity	4	Heated lid off

3.5 Purification of Enriched Library

1. Warm the bottle of Agencourt AMpure XP reagent to room temperature for 30 min. Gently vortex the bottle to resuspend any magnetic particles that may have settled.
2. Label two 1.5 mL microcentrifuge tubes for each 5 mL centrifuge tube of pooled enriched library: one for purification and one for the eluted library.
3. Transfer 700 μL of the barcoded pooled library into one of the 1.5 mL microcentrifuge tubes.
4. Add 630 μL (0.85× volumes) of Agencourt AMpure XP reagent. Mix reagent and library thoroughly by slowly pipetting up and down 20 times. Incubate for 10 min at room temperature for maximum recovery.

5. Place tube onto the Invitrogen DynaMag-2 for at least 2 min to separate beads from solution. Wait for the solution to clear, and then aspirate the cleared solution from the tube and discard.
6. With tube on the DynaMag-2 magnet, add 700 μL of 70% ethanol. Incubate for 30 s at room temperature. Aspirate the ethanol and discard. Repeat once more for two washes. Remove all ethanol without disturbing the magnetic beads (*see* **Note 14**).
7. Remove tube from the magnet.
8. Add 50 μL nuclease-free water to the tube. Resuspend the beads in the water by pipetting up and down 10 times or until the beads are completely resuspended. Incubate at room temperature for 5 min for maximum recovery.
9. Place tube onto the DynaMag-2 magnet for 2 min to separate beads from the water containing the eluted library. Wait for the solution to clear.
10. Transfer 48 μL of the eluted library into the other 1.5 mL microcentrifuge tube. Place on ice or store at −20 °C for up to 4 weeks.

3.6 Validation of Library

1. Analyze 1 μL of library using Agilent DNA 7500 Kit and Agilent Bioanalyzer 2100 Instrument operating Agilent Bioanalyzer 2100 Expert Software. Evaluate the Bioanalyzer trace to confirm the library's average fragment length. The Bioanalyzer trace should display a strong peak at 500 ± 20 bp. Trailing peaks are expected and do not interfere with Illumina sequencing or downstream analysis.
2. Estimate the library's concentration using KAPA Library Quantitation Kit for Illumina Libraries. Prepare a minimum of three independent replicates of 10,000 times diluted library in 0.1% Tween-20 solution using the real-time PCR conditions outlined in Table 12. The library's molar concentration is calculated using KAPA Quant Data Template (https://www.kapabiosystems.com/document/kapa-library-quantification-data-analysis-template/?dl=1), which reports the undiluted concentration in both ng/μL and nM. The library's molar concentration is expected to be 4.0 nM or greater.

3.7 Illumina Sequencing

Normalize the library's molar concentration to 4.0 nM in 20 μL by diluting the library with water (*see* **Note 15**). We sequenced our library using Sequencing-by-Synthesis (SBS) chemistry on the MiSeq System programmed to generate 2 × 250-bp paired-end reads using the MiSeq Reagent Kit v2 500-cycle. To achieve optimal cluster density, we recommend 7.0 pM of denature library and 1.2 pM of denatured PhiX V3 Control is loaded onto the flow cell. In real-time, the MiSeq System's Controller Software captures

Table 12
Real-time PCR amplification program

Steps	Time/cycles	Temperature (°C)
1. Denature	5 min	95
2. Denature	30 s	95
3. Annealing/extension	45 s	60
4. Repeat steps 2–3	34 cycles more	

images of the flow cell, which it interprets and converts from fluorescent intensities to base calls. Upon completion, up to 30 million paired-end reads are reported in FASTQ file format.

4 Notes

1. If iProof DNA Polymerase is unavailable, select an enzyme with these equivalent specifications:
   (a) High-fidelity *Pyrococcus*-like enzyme with processivity-enhancing domain.
   (b) Error rate of 4.4×10^{-7} in buffer provided.
   (c) Incompatible with dUTP.
   (d) Blunt end DNA products.
   (e) Magnesium dependent.
   (f) Compatible with DMSO, Formamide, glycerol, and betaine.
   (g) One unit is defined as the amount of enzyme required to incorporate 10 nmol of dNTPs into acid insoluble form at 74 °C in 30 min under the manufacturer's stated assay conditions.
2. As the density of *Wolbachia* varies among species [e.g., 14, 15] concentration of DNA might vary between organisms. Total DNA needed for six different reactions is 6 μL. We recommend 15 μL of DNA because more volume in each well makes it easier to work with multichannel pipettes and reduces pipetting error.
3. When choosing our restriction enzymes, their recognition sites must be absent from the target gene sequences and produce an amplicon with 5′ overhangs once digested. We are required to use EcoR1 restriction enzyme because the EcoR1 recognition site sequence already exists in the barcoded adapters. Replacing the EcoR1 recognition site means redesigning and purchasing

new barcoded adapters. To make our target gene compatible with the existing ddRADseq EcoRl adapters, we must add the EcoRl recognition site to our forward gene-specific primers. To our reverse gene-specific primers, we choose to add the Notl recognition site because the Notl enzyme does not cleave our target gene sequence. Once digested with both EcoRl and Notl, the amplified target gene has the necessary sticky-ends for ligation. Because the restriction enzyme will cleave at the end of the amplicon, two flanking bases (AT) are added to the 5′ end of each primer to improve cleavage efficiency. We recommend determining the number of flanking bases needed if the restriction enzymes used for the double digest are changed.

4. The T4 DNA Ligase Buffer is provided with the Ligase. Thaw buffer at room temperature to avoid precipitate formation. If present, warm to 37 °C for 2 min followed by vortexing, and repeat until the precipitate has dissolved.
5. Phosphothiolate bonds are added to the first two bases of the 5′-end of both the forward and reverse primer. These phosphothiolate bases are a variant of normal DNA; a sulfur molecule replaces one of the non-bridging oxygen molecules. Phosphothiolation inhibits endo-nuclease and exo-nuclease from acting on the adapter-ligated DNA fragments during the PCR amplification reaction.
6. The prepared Illumina PCR primer mix is enough working stock for 192 × 20 μL reactions.
7. Multiplexing the six primer pairs is inadvisable as different amplification efficiencies result in strong disequilibrium of the total number of Illumina sequencing reads per gene.
8. PCR reactions are performed separately for each gene. Therefore, optimized PCR conditions can vary between reactions.
9. We documented the amplification success of each individual reaction as strong amplification, weak amplification, or no amplification. In Subheading 3.4, **step 9**, we pooled PCR-enriched libraries based on their documented amplification success at Subheading 3.1, **step 7**. This allows for parallel purification in **step 3.5** of the gene products with varied degrees of amplification success.
10. This annealing step needs to be performed only once. The prepared EcoRl Adapter 96-well plate is enough annealed, double-stranded EcoRl adapter stock for up to 80 96-well plates. We recommend aliquoting the annealed adapter plate to minimize freeze-thaw cycles.
11. This annealing step only needs to be performed once. The Notl adapter is common to all fragments after ligation and

does not contain a unique barcode sequence. The prepared 150 μL stock is enough annealed, double-stranded NotI adapter stock for a single 96-well Adapter Ligation plate. It is easiest to have a large amount of NotI annealed adapter aliquoted in 1.5 mL tubes at −20 °C.

12. Although the barcoded EcoRI adapter plate and NotI adapter stock can be prepared at this step, we recommend preparing both before the start of the experiment (*see* **Notes 10** and **11**).
13. The product can be stored at −20 to −80 °C for up to 3 months.
14. The beads are dry when (1) there is no residual ethanol on the sides or bottom of the tube, (2) the beads are not surrounded by a droplet of liquid, and (3) they have a glossy or "wet" appearance. The presence of ethanol interferes with Illumina Sequencing. Remove as much ethanol as possible by aspiration. Allow enough time for the ethanol to evaporate; this can take up to 15 min. However, do not wait so long that the glossy or "wet" appearance becomes brittle and grainy. When this happens, the beads have over dried, and the library is difficult to elute; yields will be lower than expected.
15. To calculate the volume of stock library needed, multiply 4.0 nM by 20 μL and divide by the stock library's molar concentration as determined in Subheading 3.6, **step 2**.

Acknowledgment

We thank Melissa Stephens, Michael Pfrender, and Stuart Jones for helpful discussions. HS is supported by the Erwin Schrödinger Fellowship J-3527-B22 of the Austrian Science Fund FWF. In addition, the work was aided by support to JLF from the National Science Foundation and United States Department of Agriculture.

References

1. Hurst GDD, Frost CL (2015) Reproductive parasitism: maternally inherited symbionts in a biparental world. Cold Spring Harb Perspect Biol 7:a017699
2. Engelstädter J, Hurst GDD (2009) The ecology and evolution of microbes that manipulate host reproduction. Annu Rev Ecol Evol S 40:127–149
3. Weinert LA, Araujo-Jnr EV, Ahmed MZ, Welch JJ (2015) The incidence of bacterial endosymbionts in terrestrial arthropods. Proc R Soc B 282:20150249
4. Wu M, Sun LV, Vamathevan J et al (2004) Phylogenomics of the reproductive parasite *Wolbachia pipientis w*Mel: a streamlined genome overrun by mobile genetic elements. PLoS Biol 2:327–341
5. Klasson L, Westberg J, Sapountzis P et al (2009) The mosaic genome structure of the *Wolbachia w*Ri strain infecting *Drosophila simulans*. Proc Nat Acad Sci U S A 106:5725–5730
6. Ellegaard KM, Klasson L, Näslund K et al (2013) Comparative genomics of *Wolbachia* and the bacterial species concept. PLoS Genet 9:e1003381
7. Braig HR, Zhou WG, Dobson SL, O'Neill SL (1998) Cloning and characterization of a gene encoding the major surface protein of the bacterial endosymbiont *Wolbachia pipientis*. J Bacteriol 180:2373–2378

8. Baldo L, Dunning Hotopp JC, Jolley KA et al (2006) Multilocus sequence typing system for the endosymbiont *Wolbachia pipientis*. Appl Environ Microbiol 72:7098–7110
9. Baldo L, Ayoub NA, Hayashi CY et al (2008) Insight into the routes of *Wolbachia* invasion: high levels of horizontal transfer in the spider genus *Agelenopsis revealed* by *Wolbachia* strain and mitochondrial DNA diversity. Mol Ecol 17:557–569
10. Schuler H, Arthofer W, Riegler M et al (2011) Multiple *Wolbachia* infections in *Rhagoletis pomonella*. Entomol Exp Appl 139:138–144
11. Stahlhut JK, Desjardins CA, Clark ME et al (2010) The mushroom habitat as an ecological arena for global exchange of *Wolbachia*. Mol Ecol 19:1940–1952
12. Ahmed MZ, Breinholt JW, Kawahara AY (2016) Evidence for common horizontal transmission of *Wolbachia* among butterflies and moths. BMC Evol Biol 16:118
13. Schuler H, Bertheau C, Egan SP et al (2013) Evidence for a recent horizontal transmission and spatial spread of *Wolbachia* from endemic *Rhagoletis cerasi* (Diptera: Tephritidae) to invasive *Rhagoletis cingulata* in Europe. Mol Ecol 22:4101–4111
14. Arthofer W, Riegler M, Schneider D et al (2009) Hidden *Wolbachia* diversity in field populations of the European cherry fruit fly, *Rhagoletis cerasi* (Diptera, Tephritidae). Mol Ecol 18:3816–3830
15. Arthofer W, Riegler M, Avtzis DN, Stauffer C (2009) Evidence for low-titre infections in insect symbiosis: *Wolbachia* in the bark beetle *Pityogenes chalcographus* (Coleoptera, Scolytinae). Environ Microbiol 11:1923–1933
16. Gerth M, Bleidorn C (2016) Comparative genomics provides a timeframe for *Wolbachia* evolution and exposes a recent biotin synthesis operon transfer. Nat Microbiol 2:16241
17. Bleidorn C, Gerth M (2018) A critical re-evaluation of multilocus sequence typing (MLST) efforts in *Wolbachia*. FEMS Microbiol Ecol 94(1), fix163.
18. Parchman T, Gompert Z, Benkman C, Schilkey F, Mudge J, Buerkle CA (2012) Genome wide association mapping of an adaptive trait in lodgepole pine. Mol Ecol 21:2991–3005
19. Baird NA, Etter PD, Atwood TS et al (2008) Rapid SNP discovery and genetic mapping using sequenced RAD markers. PLoS One 3: e3376
20. Peterson BK, Weber JN, Kay EH et al (2012) Double digest RADseq: an inexpensive method for de novo SNP discovery and genotyping in model and non-model species. PLoS One 7:e37135
21. Schloss PD, Westcott SL, Ryabin T et al (2009) Introducing mothur: open-source, platform-independent, community-supported software for describing and comparing microbial communities. Appl Environ Microbiol 75:7537–7541

Chapter 15

Preparation and Use of a Yeast shRNA Delivery System for Gene Silencing in Mosquito Larvae

Keshava Mysore, Limb K. Hapairai, Na Wei, Jacob S. Realey, Nicholas D. Scheel, David W. Severson, and Molly Duman-Scheel

Abstract

The mosquito genome projects facilitated research in new facets of mosquito biology, including functional genetic studies in the dengue and Zika virus vector *Aedes aegypti* and the primary African malaria vector *Anopheles gambiae*. RNA interference (RNAi) has facilitated gene silencing experiments in both of these disease vector mosquito species and could one day be applied as a new method of vector control. Here, we describe a procedure for the genetic engineering of *Saccharomyces cerevisiae* (baker's yeast) that express short hairpin RNA (shRNA) corresponding to mosquito target genes of interest. Following cultivation, which facilitates inexpensive propagation of shRNA, the yeast is inactivated and prepared in a ready-to-use dry tablet formulation that is fed to mosquito larvae. Ingestion of the yeast tablets results in effective larval target gene silencing. This technically straightforward and affordable technique may be applicable to a wide variety of mosquito species and potentially to other arthropods that feed on yeast.

Key words *Aedes aegypti*, *Anopheles gambiae*, *Saccharomyces cerevisiae*, RNAi, Dengue, Zika, Malaria, Vector, Development, Insect

1 Introduction

Blood feeding *Aedes* and *Anopheles* vector mosquitoes transmit disease-causing pathogens that result in hundreds of thousands of human deaths each year. Dengue, a leading cause of morbidity in the tropics, Zika, a public health emergency of international concern, as well as yellow fever and chikungunya, result from infections with arboviruses transmitted through the bites of *Aedes* mosquitoes [1]. The global incidence of dengue has increased dramatically, with over 400 million estimated cases occurring each year [2]. Cases of Zika, which has been linked to severe birth defects and neurological disorders, are currently occurring in many countries in the Americas, and Zika has rapidly spread to previously

Keshava Mysore and Limb K. Hapairai contributed equally to this work.

Susan J. Brown and Michael E. Pfrender (eds.), *Insect Genomics: Methods and Protocols*, Methods in Molecular Biology, vol. 1858, https://doi.org/10.1007/978-1-4939-8775-7_15, © Springer Science+Business Media, LLC, part of Springer Nature 2019

unaffected geographic areas [3]. Malaria results from infection with *Plasmodium spp.* parasites, which are transmitted to people through the bites of infected *Anopheles* mosquitoes, including the primary African vector *Anopheles gambiae* [4]. Despite the devastating global impact of mosquito-borne illnesses on human health, effective means of preventing and treating these diseases are lacking, and mosquito control is presently the best method of disease prevention.

In recent years, advances in the genetic engineering of mosquitoes have made the potential for using transgenic vector control strategies a reality [5–7], challenging researchers to identify novel gene targets for vector control and additional methods of manipulating mosquito gene function. Altering gene expression during development, which proved useful for generation of the female-flightless control intervention [5], may promote the elucidation of novel mosquito control strategies. However, to date, the functions of very few genes have been characterized during disease vector mosquito development. RNA interference (RNAi), initially discovered in *C. elegans* [8], has facilitated characterization of gene function in a wide variety of organisms, including insects [9, 10]. The RNAi pathway is initiated by Dicer, which cleaves long dsRNA into short 21–25 nucleotide-long small interfering RNAs (siRNAs) that function as sequence-specific interfering RNA molecules. siRNAs silence genes that are complementary in sequence by promoting transcript turnover, cleavage, and disruption of translation [10]. Although most mosquito researchers use longer (300–400 bp) dsRNA molecules for RNAi experiments, the short length of custom siRNAs and their short hairpin RNA (shRNA) counterparts facilitates design of interfering RNA with less potential for off-site targeting. It is also possible to confirm gene silencing phenotypes by performing experiments with multiple siRNAs that recognize different target sites within a gene of interest. Moreover, if siRNAs were to one day be used as insecticides, the development of multiple siRNA insecticides to silence the targeted gene will be useful for combating resistance resulting from a point mutation in any single target site. Additionally, the use of short sequences facilitates the design of interfering RNA molecules that recognize target sites that are not found in nontarget organisms, but which are conserved in multiple mosquito species.

Although RNAi does not generate heritable germline mutations, it offers several advantages that may be of utility. First, through management of the timing of siRNA/shRNA delivery, researchers can control the time at which gene silencing initiates. This advantage can be used to overcome challenges such as developmental lethality or sterility, issues which can hinder both the production and maintenance of strains bearing heritable mutations. Moreover, genetic engineering of non-model insects is still a relatively expensive and labor-intensive process. Thus, although the degree of gene silencing by RNAi can vary depending on the gene targeted, the tissue type, and also from subject to subject, RNAi is still frequently used for functional genetics studies in mosquitoes and other insects [9, 10]. Several different interfering RNA delivery strategies have been implemented in developing mosquitoes. For example, we have successfully used microinjection to deliver siRNAs for silencing of developmental genes in *A. aegypti* embryos, larvae, and pupae [11–18]. However, this labor-intensive delivery strategy, which requires both technical skill and a microinjection setup, cannot be extended to the field. Although ingestion-based strategies do not work in all insect species, notably *Drosophila melanogaster*, RNAi through oral delivery promotes gene silencing in a variety of different insects, including mosquitoes [9, 19–24]. For example, soaking mosquitoes in interfering RNA solution, which they ingest, can induce gene silencing [19, 20], and this convenient technique requires little equipment or labor. Chitosan nanoparticle-mediated interfering RNA delivery also effectively silences genes in *A. gambiae* [22] and *A. aegypti* larvae [23, 24]. However, while soaking and chitosan/siRNA methodology facilitate relatively affordable laboratory studies and require little equipment or labor [15], the present costs of RNA synthesis may still be a concern in large-scale laboratory and field applications.

The use of microbes facilitates affordable RNA propagation through microbe cultivation. For example, *Escherichia coli* strains engineered to produce double stranded (dsRNA) molecules targeting several genes of interest have generated successful gene silencing when fed to *A. aegypti* [25, 26] and *A. gambiae* [27] larvae. We have also engineered *Saccharomyces cerevisiae* (baker's yeast), a model organism that is genetically tractable and inexpensive to culture, to produce shRNA corresponding to target sequences in several mosquito genes of interest [26, 27]. Yeast, which is added to chitosan nanoparticles [15] for larval feeding assays, is a strong odorant attractant and a source of nutrition for laboratory-bred mosquito larvae. Therefore, *S. cerevisiae,* unlike *E. coli* or chitosan, can serve both as the RNA delivery system and directly as the bait. Moreover, if RNAi were to one day be introduced as a mechanism for mosquito control, it seems likely that yeast, a natural product that is used in food and beverage production across the globe, could be a microbe that is readily accepted by users and stakeholders. Our laboratory studies demonstrate that yeast interfering RNA strains, which can function as larvicides, effectively silence genes in both *Aedes* [26] and *Anopheles* [27] larvae. Moreover, we have developed a procedure for production of ready-to-use dried and inactivated yeast interfering RNA tablets, a formulation which facilitates laboratory studies and that could one day promote the seamless integration of these larvicides into mosquito control programs. Importantly, laboratory trials demonstrate that the larvicidal capacity of this yeast is maintained even when the microbes are heat-killed [26, 27], an important advantage given that additional regulatory approvals are often required for the feeding of living recombinant organisms to live animals in the laboratory, as well as in the field. *S. cerevisiae* has been cultivated worldwide for thousands of years, and this technology, which is readily scaled, can be adapted in most laboratories using the methodology for recombinant yeast production (Subheadings 3.1 and 3.2 below), yeast culturing (Subheading 3.3) and tablet preparation (Subheading 3.4), larval feeding assays (Subheading 3.5), and confirmation of gene silencing (Subheadings 3.6 and 3.7) described herein.

2 Materials

2.1 Equipment

1. Horizontal gel electrophoresis system.
2. Microwave oven.
3. Gel imaging system.
4. Nanodrop or comparable spectrophotometer.
5. Shaker incubator.
6. Microcentrifuge (i.e., Eppendorf 5415D or a comparable model).
7. High speed refrigerated centrifuge (i.e., Eppendorf 5810R plus or a comparable model).
8. Water bath (a shaking bath is preferred, though not required for Subheading 3.7).
9. Heat block.
10. Vortex.
11. Incubator with good outward ventilation.
12. Micropipetters.
13. Ice bucket.
14. Racks for holding microfuge and centrifuge tubes.
15. Dissection microscope.
16. qRT-PCR machine (if detecting silencing via qRT-PCR, Subheading 3.6).
17. Rocking platform shaker or nutator (if detecting silencing via in situ hybridization, Subheading 3.7).
18. Cell spreader.
19. Bunsen burner.

2.2 Construction of shRNA Expression Vector

Molecular grade reagents and ultrapure water should be used to prepare sterile solutions and media in Subheadings 2.2–2.6.

1. Shuttle vector: Use of the *pRS426 GPD* yeast shuttle vector permits constitutive expression of inserts cloned into the multiple cloning site downstream of a *GPD* promoter and upstream of a *cyc1* terminator [28]. *pRS426* has an ampicillin resistance marker for selection in *E. coli*, in which all cloning experiments are performed. The *pRS426 URA3* marker facilitates selection of *S. cerevisiae* that have been successfully transformed with the construct.
2. Custom oligonucleotides encoding the shRNA transcript; store annealed oligonucleotides at −20 °C.
3. T4 DNA ligase and buffer stored at −20 °C.
4. LB media: 10 g tryptone, 5 g yeast extract, 10 g NaCl (add 15 g of bacto-agar for LB plates). Bring the volume to 1000 mL with ultrapure water, pH to 7.5, then autoclave and store at 4 °C. *E. coli* DH5α competent cells transformed with the *pRS426* shRNA expression plasmid are grown at 37 °C on standard LB media containing 100 μg/mL ampicillin.
5. Plastics: Sterile Petri dishes, 1.5 mL microfuge tubes, 10 mL culture tubes.
6. Plasmid miniprep kit.
7. Restriction enzymes corresponding to the *pRS426 GPD* multiple cloning site (stored at −20 °C); we typically use the BamHI (5′) and XhoI (3′) sites.
8. Agarose (stored at room temperature).
9. 50× TAE buffer: 242.2 g Tris base, 57.1 mL glacial acetic acid, 18.6 g EDTA, and ultrapure water to volume of 1 L with ultrapure water. Stored at room temperature and diluted 1:50 with ultrapure water for 1× working concentration.
10. Ethidium bromide (10 mg/mL in ultrapure water). Stored at 4 °C and diluted 1:20 for gel electrophoresis.
11. Sterile 500 mL Erlenmeyer flask.
12. Sequencing primers M13F 5′GTAAAACGACGGCCAGT3′ and M13R 5′CACACAGGAAACAGCTATGACCAT3′; store at −20 °C.

2.3 Yeast Transformation, Culturing, and Preparation of Dried Inactivated Tablets

1. *S. cerevisiae* strain *BY4742 (genotype MATα his3Δ1 leu2Δ0 lys2Δ0 ura3Δ0)* [29].
2. 10× Lithium acetate (1 M in ultrapure water). Filter sterilized and stored at 4 °C.
3. 50% PEG 3350 in ultrapure water. Filter sterilized, aliquoted, wrapped in parafilm, and stored at 4 °C.

4. Salmon sperm DNA (2 mg/mL in ultrapure water); should be phenol chloroform extracted, sheared, and stored at −20 °C. Heat-denature prior to use in transformation.
5. Standard YPD media: 20 g bacto peptone, 10 g yeast extract, and water to 900 mL volume. The solution should be autoclaved, then cooled to the touch prior to the addition of 100 mL sterile 20% glucose solution and storage at 4 °C.
6. Petri dishes (sterile).
7. Synthetic complete media with dextrose lacking uracil (SCD-ura): 1.7 g of yeast nitrogen base, 5 g ammonium sulfate, and 0.77 g complete supplement mixture lacking uracil (CSM-ura). For plates, add 20 g of agar. Bring the volume to 900 mL with ultrapure water and autoclave. Once the media has cooled to the touch, add 100 mL of sterile 20% glucose solution to the cooled media. Store media plates at 4 °C for up to 1 month.
8. Tubes: Sterile 1.5 and 2 mL microfuge tubes, 10 mL culture tubes, and 50 mL centrifuge tubes.
9. Sterile 1 L Erlenmeyer flasks.

2.4 Mosquito Strains, Rearing, and Recombinant Yeast Feeding Assays

1. Maintain *A. gambiae* M strain or other strains of interest according to standard lab practice or as described [30, 31].
2. Maintain *A. aegypti* Liverpool IB12 or other strains of interest according to standard lab practice or as previously described [31].
3. Membrane Feeding System (i.e., Hemotek Ltd.; replaces vertebrate animals for blood feeding mosquitoes) and commercially purchased blood (i.e., defibrinated sheep's blood; store at 4 °C).
4. 500 mL plastic cups and tray in which to place the cups.
5. Petri dish lid or mesh to cover cups (holes should be small enough to prevent adult escape).
6. Rubber bands to secure mesh to cups.

2.5 Detection of Target Gene Silencing by qRT-PCR

1. Trizol reagent (Invitrogen); stored at 4 °C.
2. High capacity RNA to cDNA Kit (i.e., Applied Biosystems kit); stored at −20 °C.
3. SYBR green I PCR kit (i.e., Applied Biosystems kit); stored at 4 °C.
4. Sterile PCR tubes.

2.6 Detection of Target Gene Silencing by Whole Mount In Situ Hybridization

1. PEM: 34.63 g of PIPES disodium salt, 10 mL of 0.2 M EGTA, and 1 mL of 1 M $MgSO_4$. Add ultrapure water and mix for 20 min, then adjust pH to 7.0 with HCl. Bring to 1 L volume and filter through a 0.22 μm filter. Store this solution at 4 °C

for up to 1 year. To prepare PEM-F, dilute 37% formaldehyde 1:10 in PEM just prior to use.

2. DIG-labeled riboprobe prepared with DIG Labeling Kit (SP6/T7, Roche Life Science); stored at −20 °C.
3. Hybridization solution (Hyb): 25 mL deionoized formamide, 12.5 mL 20× SSC, 2.5 mg heparin, 50 μL Tween-20. Adjust the pH to 5.0 using HCl. Bring the final volume to 50 mL with sterile ultrapure water and store at −20 °C.
4. Hyb-DNA-SDS: 14.85 mL Hyb solution with 150 μL of 10 mg/mL sonicated salmon sperm DNA. Store at −20 °C; just prior to use, warm the Hyb-DNA solution and add SDS to a final concentration of 0.3%.
5. 10× PBS: 11.9 g Na_2HPO_4, 2.23 g NaH_2PO_4 (anhydrous), and 102.2 g NaCl. Bring to 1 L total volume with ultrapure water and the pH adjusted to 7.4. Autoclave and store at room temperature. Dilute to 1× working concentration prior to use.
6. PT: PBS and 0.1% Triton X-100 stored at room temperature.
7. PTw: PBS and 0.1% Tween-20 stored at room temperature.
8. Detergent solution: 5 mL 10% SDS, 250 μL Tween-20, 5 mL 0.5 M Tris–HCl (pH 7.5), 100 μL 0.5 M EDTA (pH 8), and 7.5 mL 1 M NaCl. Add ultrapure water to 50 mL and store at room temperature.
9. AP buffer and color reaction solution: 250 μL 1 M $MgCl_2$, 5 mL 1 M NaCl, 5 mL 1 M Tris (pH 9.5), 50 μL Tween-20, and ultrapure water to 50 mL. Prepare fresh at room temperature just prior to use. For the color reaction, add 20 μL NBT-BCIP solution (Roche) to 1 mL of AP buffer and use immediately.
10. Glycerol solution: 50% and 70% glycerol solutions are prepared with ultrapure glycerol in PBS. Store at room temperature.

3 Methods

3.1 Preparation of Yeast Transformation Vector

1. Select shRNA target sites and design shRNAs according to standard lab practice. Scramble the sequence of a knockdown shRNA target site to design negative control shRNA that does not correspond to any mosquito gene. The shRNA sequence designer tool [32] will facilitate design of DNA oligonucleotides corresponding to each target sequence and allows the user to add restriction enzyme overhangs to facilitate downstream cloning (*see* **Note 1**). DNA oligonucleotides corresponding to the custom shRNA sequences can be purchased through a number of reputable vendors.

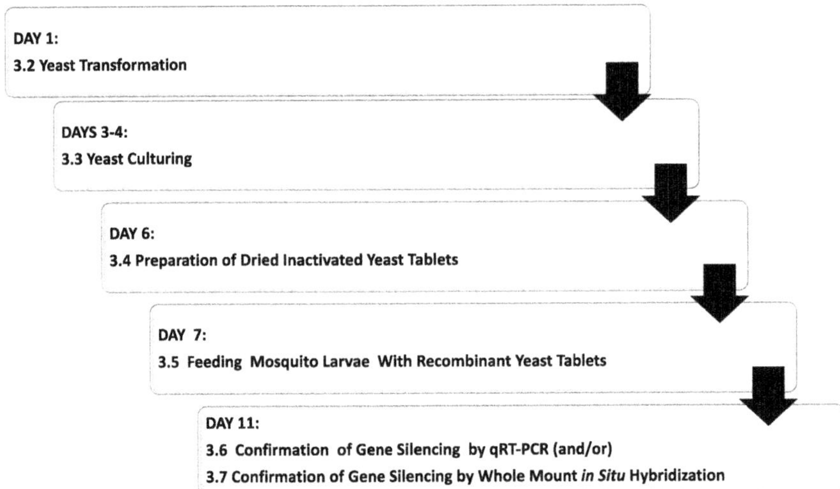

Fig. 1 Experimental timeline. The sequence of experimental events over an ~11 day experimental timeline is presented for **steps 3.2** through **3.7**, which initiate following preparation of the shRNA expression construct (**step 3.1**) and conclude with analysis of silencing in fourth instar larvae

2. DNA oligonucleotides corresponding to the shRNA sequence are annealed and cloned into the multiple cloning site of the *pRS426* shuttle vector [28] (*see* **Note 2**) using standard double digestion and cloning techniques (*see* **Notes 3** and **4**). Confirm positive colonies by harvesting and double digestion of the plasmid DNA followed by agarose gel electrophoresis to confirm inserts. The correct sequences of positive inserts should be verified through sequencing with the M13F or M13R primer (*see* **Note 5**).

3.2 Yeast Transformation

A timeline outlining the sequencing of Subheadings 3.2 through 3.7 is presented in Fig. 1.

1. Inoculate 2–5 mL of liquid YPD with *S. cerevisiae* strain *BY4742* and incubate with shaking overnight at 30 °C and 250 rpm.
2. Use 600 μL of the pre-culture (which should be at an OD_{600} of ~0.2) to inoculate 25 mL of fresh YPD. Shake at 250 rpm and 30 °C until the OD_{600} reaches ~0.8 (approximately 3–5 h).
3. Harvest the culture in a sterile 50 mL centrifuge tube spun at 1000 × *g* for 5 min.
4. Pour off the medium and resuspend the cells in 25 mL of sterile water. Centrifuge again at 1000 × *g* for 5 min.
5. Pour off the water and resuspend the cells in 1.0 mL of 0.1 M (1×) LiAc. Transfer the cell suspension to a 1.5 mL microfuge tube.

6. Pellet the cells at 8000 rpm for 30 s and remove the LiAc with a micropipette.
7. Resuspend the cells to a final volume of ~250 μL by adding 200 μL of 0.1 M (1×) LiAc (2 × 10^9 cells/mL).
8. Boil the salmon sperm DNA for 5 min and quickly chill it on ice (*see* **Note 6**).
9. Gently mix the cell suspension and pipette 50 μL samples into microfuge tubes. Pellet the cells and remove the LiAc with a micropipette.
10. Prepare the transformation mixture, which consists of 240 μL PEG (50% w/v) (*see* **Note 7**), 36 μL 1.0 M LiAc, 50 μL heat-denatured salmon sperm DNA (2.0 mg/mL), ~3.0 μg of plasmid DNA, and sterile ultrapure water to a total volume of 360 μL.
11. Add the transformation mixture to the cell pellet. Vortex each tube briefly until the cells are resuspended.
12. Heat shock the cells in a water bath at 42 °C for 15–20 min.
13. Microfuge at 6000–8000 rpm for 30 s and remove the supernatant with a micropipette.
14. Pipette 600 μL of sterile water into the tube and resuspend the pellet by pipetting it up and down gently.
15. Plate the cells on SCD-ura media plates.
16. Incubate the plates for 2–4 days at 30 °C to recover transformants. Re-streak yeast from an individual clone onto a new SCD-ura media plate (*see* **Note 8**).

3.3 Yeast Culturing and Preparation of Dried Inactivated Yeast Tablets

1. To prepare yeast from each desired strain (experimental or control), inoculate a 2 mL culture of yeast SCD-ura media placed in a sterile 10 mL culture tube (*see* **Notes 9** and **10**). Allow the culture to grow for 24 h in a shaker incubator set at 30 °C and 250 rpm. The culture tube should be placed at a 45° angle.
2. Transfer the 2 mL culture into a 1 L flask with 250 mL SCD-ura media and allow it to grow at 30 °C and 250 rpm in a shaker incubator until the OD^{600} of the culture is between 2.5 and 3.0 (*see* **Notes 11** and **12**).

3.4 Preparation of Dried Inactivated Yeast Tablets

1. Transfer the 250 mL culture into five 50 mL centrifuge tubes, each with 50 mL of culture. Spin the balanced tubes for 20 min at 4000 rpm (*see* **Note 13**). Decant the supernatant.
2. Heat kill the yeast by incubating the tubes containing the pellets in a water bath set to 70 °C for 5 min; once heat-killed, transfer (*see* **Note 14**) the pellets into 2 mL Eppendorf tubes (*see* **Note 15**).

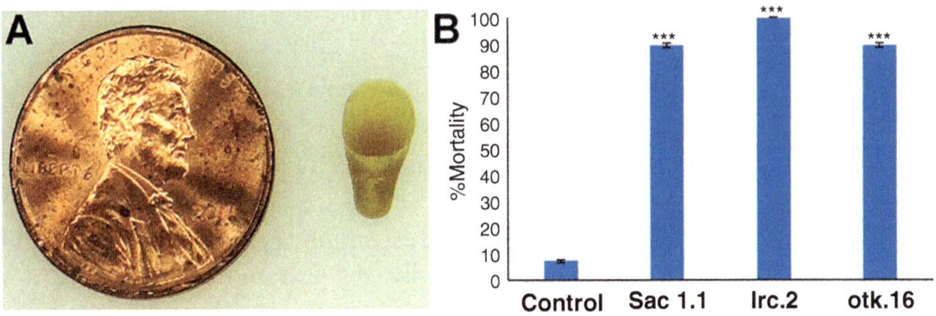

Fig. 2 Yeast interfering RNA tablets induce significant *A. gambiae* larval death. Dried inactivated yeast interfering RNA tablets (**A**; penny shown for scale) were prepared and fed to 20 *A. gambiae* larvae. Significant death was observed in larvae fed with yeast expressing shRNA hairpins corresponding to the *Sac1, lrc,* and *otk* genes as compared to larvae fed control yeast interfering RNA tablets (**B**). These data were compiled from three biological replicate experiments ($n = 240$ larvae total/condition) and analyzed by ANOVA with Tukey's multiple comparison test. ***$p < 0.001$ as compared to control-fed larvae; error bars denote standard error of the mean (SEM). Reproduced through open access from ref. 27

3. Spin the 2 mL tubes at ~13,000 rpm for 2 min and remove the supernatant entirely by pipetting. Place these pellets in a 30 °C incubator that has good outward ventilation (i.e., it must constantly suck moist air from the chamber). Alternatively, a food desiccator can be used for the drying process. Dry the pellets for approximately 48 h, when a gap between the pellet and the side of the tube is noted and the pellet can be dislodged from the tube (*see* **Note 16**). The final weight of each yeast tablet (shown in Fig. 2a) averages ~85 mg (~1.6 × 10^10 cells). For best results, proceed immediately to the feeding assays (Subheading 3.5). If necessary, the pellets can be stored at −80° or −20° C for 1–2 weeks.

3.5 Feeding Mosquito Larvae with Recombinant Yeast Tablets

1. Prepare 16 oz. plastic cups with 50 mL of sterile ultrapure water and label each cup appropriately. For each biological replicate experiment, prepare at least three cups per strain of yeast used (*see* **Note 17**).
2. Place one control or experimental 85 mg yeast pellet in each cup.
3. Add 20 freshly hatched age-synchronized mosquito larvae (*see* **Note 18**) to each of these cups (Fig. 3). Cover each cup with mesh secured with a rubber band or with the lid of a Petri dish. Put the cups in a plastic tray to catch any spills and place the tray in an insectary maintained at 26.5° C with relative humidity of ~85% and a light-dark cycle of 12 h. If an insectary is not available, then an incubator can be used.
4. Examine the transcript levels as discussed below and evaluate other phenotypic changes at the desired developmental time points. For example, when evaluating yeast interfering RNA

Fig. 3 Mosquito larval oral feeding assays. Larvae placed in a beaker consume yeast interfering RNA tablets. This procedure can be used to assay the impact of gene silencing on various larval phenotypes, including larval death

larvicides, larval death can be monitored (Figs. 2b and 3). For analysis of some phenotypes, it may be useful to supplement the fourth instar (L4) larval diet with an additional source of nutrition, which will allow the larvae to progress to the pupal stage more quickly (*see* **Note 19**). If adult phenotypes are to be assessed, be sure to move the cups into an adult cage.

5. It is recommended that at least three biological replicate feeding experiments are performed for gene silencing quantification and for analysis of each phenotype to be assessed.

3.6 Confirmation of Gene Silencing by qRT-PCR

1. Gene silencing can be confirmed with qRT-PCR, which can be performed according to your standard laboratory procedure or using the method summarized below, which has worked in our hands [11, 22, 23]. Perform and analyze qRT-PCR assays with at least three yeast feeding biological replicates, each with at least 10 pooled control vs. experimental larvae (we typically assess L4 animals) as described [11, 22, 23]. For analysis of a particular tissue type/body part (i.e., the brain), perform qRT-PCR following dissection to recover the tissue of interest (*see* Mysore et al. [24] for example and *see* **Note 20**).

2. Extract total RNA with Trizol reagent according to the manufacturer's instructions. Use the RNA to prepare cDNA with the High Capacity RNA to cDNA Kit (applied Biosystems) following directions supplied by the manufacturer.
3. Perform qRT-PCR using the SYBR green I PCR kit according to the manufacturer's instructions in conjunction with an Applied BioSystems Step One Plus Real-Time PCR System or a comparable model. For each biological replicate experiment, all PCR assays should be run in triplicate. Quantification of results is performed through comparison to reference standard transcript levels [11]. Transcript quantification data can be statistically analyzed using a *t*-test.

3.7 Confirmation of Gene Silencing by Whole Mount In Situ Hybridization

1. In situ hybridization permits spatial confirmation of silencing in specific tissue or cell types. Representative results are shown in Fig. 4A1, A2, B1, B2, and C1, C2). Assess tissues prepared from 20 animals (we typically use L4 animals) in at least three biological replicate yeast feeding experiments. Execute in situ hybridization according to standard lab practice or in accordance with the Haugen et al. [33] protocol, which is summarized below.
2. Synthesize digoxigenin-labeled antisense and sense control riboprobes according to standard lab practice or using the DIG RNA labeling and detection kit according to the manufacturer's directions, which are explained in further detail by Patel [34] (*see* **Note 21**).
3. A detailed protocol for tissue fixation and preparation is included in Clemons et al. [31]. In summary, larval tissues of interest are dissected and then fixed for 20 min in PEM-F (*see* **Note 22**).
4. Fixed tissues are then processed for detection of mRNA transcripts according to the method below, which was described previously in greater detail [33]. The fixed tissues can remain in eppendorf tubes throughout the procedure. Proceed with a 5 min rinse in PBS and 2× 10 min rinses in PTw. Use 1 mL wash volumes unless otherwise indicated.
5. After removing the PTw, add 1 mL of detergent solution (*see* **Note 23**) and incubate the tissues for 30 min with gentle rocking. After 30 min, remove the detergent solution and rinse the tissues 2× 10 min in PTw.
6. After removing the PTw, rinse the tissues 1× 5 min with 500 μL of 50% PTw-50% Hyb solution. Then replace the 50% PTw-50% Hyb with 500 μL of Hyb solution and incubate for 10 min. While the tissues are incubating in Hyb, boil an aliquot of the Hyb-DNA-SDS solution for 10 min in a heat block or

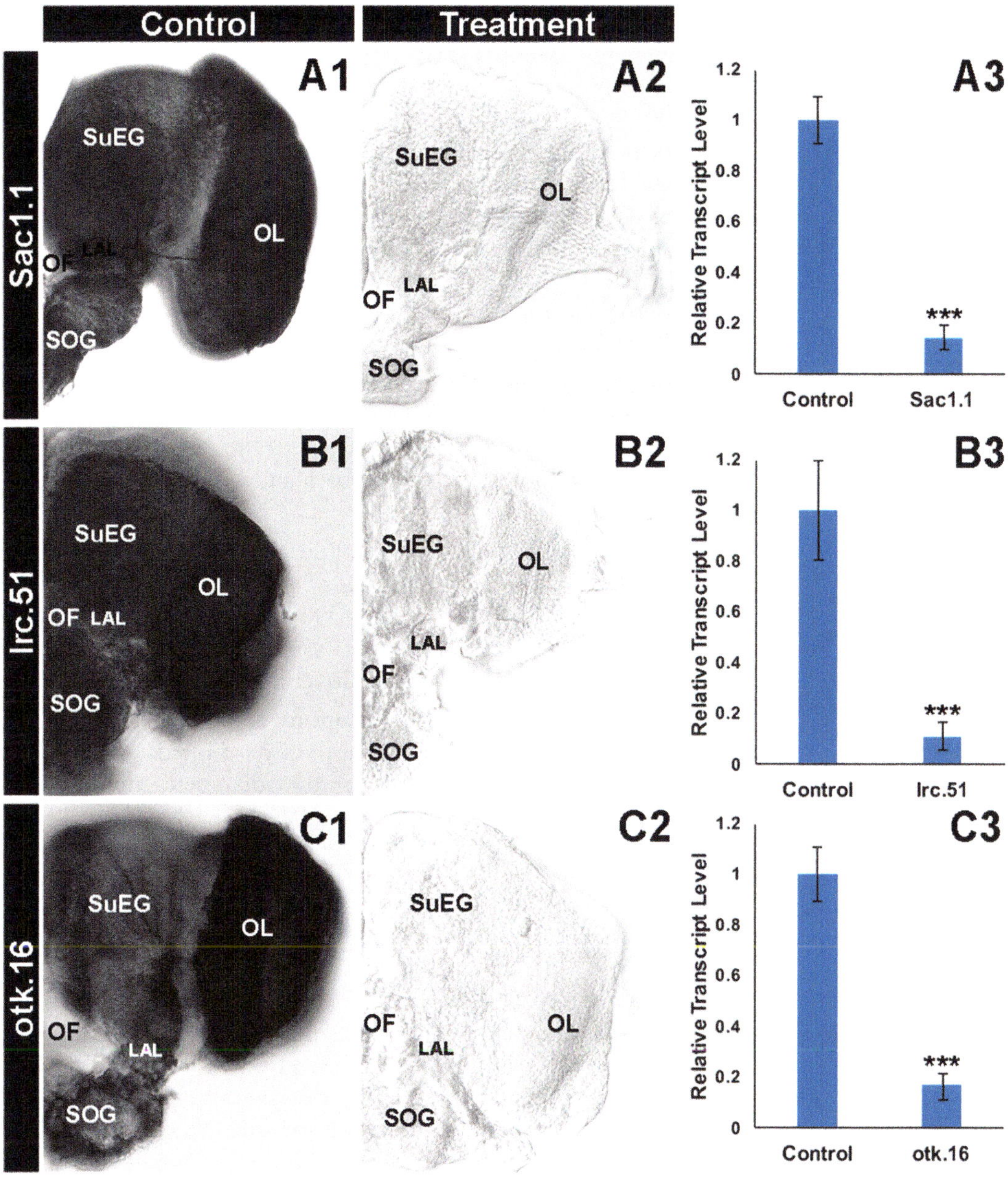

Fig. 4 Confirmed silencing of the *Sac1, Irc,* and *otk* genes in the *A. gambiae* larval brain by dried inactivated yeast interfering RNA tablets. Significantly lower *Sac1* (A1–A3), *Irc* (B1–B3), and *otk* (C1–C3) transcript levels were detected through in situ hybridization in the L4 brains of larvae fed dried inactivated yeast interfering RNA tablets corresponding to the *Sac1* (A1), *Irc* (B1), and *otk* (C1) genes vs. animals fed with control yeast interfering RNA tablets (A2, B2, C2). For each probe, results from three biological replicate experiments were compiled ($n = 85$ total brains from larvae treated with the *Sac1* interfering RNA tablet, $n = 80$ total brains from larvae treated with the *Irc* interfering RNA tablet, and $n = 80$ brains from larvae treated with the *otk* interfering RNA tablets; $n = 40$ brains from control-treated larvae/per experiment). Data were evaluated by the Student's *t*-test. All brains are oriented dorsal upward in this figure. *LAL* larval antennal lobe, *OF* Olfactory foramen, *OL* optic lobe, *SOG* subesophageal ganglion, *SuEG* Supraesophageal ganglion. Reproduced through open access from ref. 27

boiling bath to denature the DNA; store this solution on ice while initiating **step 7**.

7. Following removal of the Hyb solution from the tissues, add 100 μL of denatured Hyb-DNA-SDS solution to the tissues and place the tubes in a floating rack located in a 60 °C water bath for 60 min. A shaking bath is preferred, though not absolutely required, for blocking, overnight hybridization, and post-hybridization washes. Toward the end of the 60 min incubation, heat denature the riboprobe suspended in Hyb-DNA-SDS solution by boiling it for 5 min in a heat block or boiling bath; store briefly on ice while initiating the next step.
8. Remove the Hyb-DNA-SDS blocking solution from the tissues and replace it with the riboprobe/Hyb-DNA-SDS solution. Gently stir the mixture with a pipette tip, and then place the tubes back in the 60 °C water bath and continue with hybridization overnight.
9. In the morning, perform the following wash steps in the 60 °C water bath using 1 mL volumes of solutions prewarmed in the 60 °C bath: 1× 30 min with Hyb solution; 5× 30 min with PTw. After the last PTw wash, bring the tubes to room temperature and wash for an additional 30 min with PT.
10. Remove the PT wash solution and add 300 μL of AP-conjugated anti-DIG antibody diluted 1:2000 in PT. Gently stir the mixture with your pipette tip and then incubate overnight at 4 °C.
11. The next day, rinse the tissue 4× 30 min with PT at room temperature.
12. After removing the PT, rinse the tissue 3× 5 min in AP buffer (*see* **Note 24**).
13. Remove the AP wash buffer and add 300 μL of AP-NBT-BCIP solution. The color reaction should be performed in the dark, monitoring the progress of the reaction by occasional visualization of the tissues with a dissection microscope. The reaction can take anywhere from several minutes to several hours to complete.
14. Stop the progress of the color reaction by removing the AP-NBT-BCIP and rinsing the tissues 4× 15 min with PT.
15. Remove the PT wash solution and rinse the tissues 1× 5 min with PBS. Following removal of the PBS solution, add 500 μL of 50% glycerol, which can be replaced after 60 min with 70% glycerol. Clear the tissues at room temperature overnight prior to mounting and analyzing the tissues in 70% glycerol. Silencing can be evaluated through transcript quantification analyses, which are performed through analysis of mean gray values (average signal intensity over the selected area) calculated for

digoxigenin-labeled transcript signal in control or experimental brains [26, 27] (Fig. 4A3, B3, C3). These analyses can be performed with Adobe Photoshop or comparable image processing software. Results should be combined from at least three biological replicate experiments, each with tissues prepared from 10 to 20 animals. Transcript quantification data can be statistically analyzed using a *t*-test.

4 Notes

1. We use the 5′ BamHI and 3′ XhoI sites in the in the *pRS426* vector and include sticky ends for cloning into these sites. Before using these sites, be sure to check that there are no internal cut sites for these enzymes in your shRNA expression insert.
2. If longer dsRNA molecules (200–400 bp) are preferred, then it is likely that the dsRNA expression system described by Murphy et al. [35] would be compatible with the yeast cultivation and mosquito oral feeding protocols described herein. However, for reasons outlined in the Introduction, we have opted to use shRNA in our own studies.
3. Although not necessary for most laboratory applications, one can also insert shRNA expression cassettes into vectors that are compatible with integration of the cassettes into the yeast genome. We have pursued this for strains that will be used in field trials [26], as it dispenses with the use of plasmids with antibiotic resistance markers and reduces the risk of horizontal transfer to other species.
4. Likewise, although not necessary for most laboratory applications, shRNA expression constructs can be placed under the control of an inducible promoter (i.e., *Gal1*), which allows for high levels of shRNA expression after integration of the expression cassettes into the yeast genome [26].
5. We have found it challenging to clone shRNA expression cassettes, presumably because a DNA oligonucleotide may bind to itself rather than to the complementary strand when the oligos are annealed prior to cloning. Likewise, although sequencing of the shRNA expression cassette clones is critical, it can also be challenging to sequence through hairpin-encoding sequence due to complications with secondary structure. We have combatted both these issues by testing multiple clones (sometimes 20 or more are necessary).
6. Although *E. coli* is readily transformed with double stranded plasmid DNA, yeast requires single stranded carrier DNA to enhance plasmid uptake.

7. PEG is prone to evaporation, which can alter the PEG:water ratio and lead to poor transformation efficiency. It may be helpful to make the PEG solution fresh for each transformation or to make it in small aliquots that are sealed tightly with parafilm.
8. Frozen glycerol stocks should be prepared from each transformant yeast strain. If you will be working with the same clones over an extended period, then it is helpful to prepare two frozen samples from each strain, one that researchers will open more frequently, and one that is set aside for long-term permanent storage. It is also useful to store the DNA plasmid in case the yeast needs to be re-transformed.
9. For the best results, use yeast that has been plated from a frozen stock on SCD-ura media plates (poured generously so that the media doesn't dry and crack) within the past week.
10. If you initiate the pre-culture in the late afternoon, it will be ready at approximately the same time the following day.
11. The culture will likely need to be diluted 1:5 using SCD-ura medium prior to taking the OD_{600} reading. The culture will likely need to grow for 42–48 h to reach an appropriate OD_{600} reading.
12. If the OD_{600} reading is less than 2.5, then let the culture grow longer. Do not harvest the culture too early.
13. If you are using a centrifuge that can spin at a higher speed, then it may be possible to reduce the length of this step.
14. It is important to use a 2 mL eppendorf tube rather than a 1.5 mL tube at this step. The 2 mL tube has a larger conical surface area that permits better drying of the yeast.
15. It is useful to perform this step in several batches, as it is easiest to transfer a warmed pellet.
16. It is important to dry the yeast thoroughly. When the yeast pellet is under-dried, it can become sticky, which makes it hard to remove from the tube and more difficult to use in feeding assays. Failure to dry the pellet thoroughly may also decrease the shelf life of the yeast interfering RNA.
17. You may wish to plan your experiments so that the larvae do not pupariate over the weekend. In our insectary, this means that we would begin experiments on a Tuesday or a Wednesday.
18. For best results, hatch *Aedes* eggs in deoxygenated water, which facilitates synchronized hatching of diapaused eggs. To deoxygenate the water, it can be autoclaved, or the eggs can be hatched in water placed under vacuum pressure. *A. aegypti*

eggs that are less than 1 month old typically hatch within 30 min at 25 psi.

19. For *Aedes*, during L4, we typically add 100 μL of a 60 g/L suspension of bovine liver powder in sterile ultrapure water (store at 4 °C). For *A. gambiae*, during L4, we have sprinkled ~10 mg of a 1:1 mixture of liver powder and nutritional yeast at the water surface. Other larval foods can likely be substituted depending on the preferred laboratory rearing protocol.
20. Examination of gene expression can be complicated because larvae will grow at different rates depending on how much food is eaten by each individual animal, and expression of many developmental genes is very dynamic. For some developmental genes, we find that expression levels differ widely even in individual control or untreated animals that were age-synchronized at the beginning of the experiment. Preparation of a specific tissue rather than use of whole animals may improve results and can be worth the effort. For some larval transcripts, we have found that in situ hybridization experiments can detect significant levels of silencing which correspond to strong and penetrant phenotypes, but that qRT-PCR experiments have at times failed to detect.
21. If multiple transcripts are expressed, be sure to synthesize a probe that corresponds to the transcript(s) of interest.
22. For some tissues, modification of the fixation solution and an overnight fixation procedure will yield better results. For example, we fix larval brains overnight at 4 °C in 1× PBS containing 1% Triton-X and 7.4% formaldehyde.
23. This SDS detergent step substitutes for the proteinase K treatment steps found in many in situ hybridization protocols and gives more consistent results than enzymatic treatment of tissues for permeabilization.
24. It is important to perform this step as described. We have noted that rinsing the tissues too quickly can significantly lengthen color reaction times.

Acknowledgments

We thank Ping Li, Longhua Sun, Chien-Wei Chen, Yingying Chen, Joanne Cunningham, and Diane Lovin for their technical assistance during development of these methodologies. Development of this protocol was funded by a United States Agency for International Development Award AID-OAA-F-16-00097 to MDS and by NIH/NIAID Award 1 R21 AI128116-01 to MDS, DWS, and NW.

References

1. CDC (2016) Surveillance and control of *Aedes aegypti* and *Aedes albopictus*. http://www.cdc.gov/chikungunya/resources/vector-control.html. Accessed May 2016
2. CDC (2018) Dengue. https://www.cdc.gov/dengue/index.html. Accessed March 2018
3. CDC (2018) Zika virus. https://www.cdc.gov/zika/index.html. Accessed February 2018
4. WHO (2016) World malaria report 2016. World Health Organization, Geneva
5. Fu G, Lees RS, Nimmo D, Aw D, Jin L, Gray P, Berendonk TU, White-Cooper H, Scaife S, Kim Phuc H, Marinotti O, Jasinskiene N, James AA, Alphey L (2010) Female-specific flightless phenotype for mosquito control. Proc Natl Acad Sci U S A 107 (10):4550–4554. https://doi.org/10.1073/pnas.1000251107
6. Wise de Valdez MR, Nimmo D, Betz J, Gong HF, James AA, Alphey L, Black WC (2011) Genetic elimination of dengue vector mosquitoes. Proc Natl Acad Sci U S A 108 (12):4772–4775. https://doi.org/10.1073/pnas.1019295108
7. Harris AF, McKemey AR, Nimmo D, Curtis Z, Black I, Morgan SA, Oviedo MN, Lacroix R, Naish N, Morrison NI, Collado A, Stevenson J, Scaife S, Dafa'alla T, Fu G, Phillips C, Miles A, Raduan N, Kelly N, Beech C, Donnelly CA, Petrie WD, Alphey L (2012) Successful suppression of a field mosquito population by sustained release of engineered male mosquitoes. Nat Biotechnol 30(9):828–830. https://doi.org/10.1038/nbt.2350
8. Fire A, Xu S, Montgomery MK, Kostas SA, Driver SE, Mello CC (1998) Potent and specific genetic interference by double-stranded RNA in *Caenorhabditis elegans*. Nature 391 (6669):806–811. https://doi.org/10.1038/35888
9. Yu N, Christiaens O, Liu J, Niu J, Cappelle K, Caccia S, Huvenne H, Smagghe G (2013) Delivery of dsRNA for RNAi in insects: an overview and future directions. Insect Sci 20 (1):4–14. https://doi.org/10.1111/j.1744-7917.2012.01534.x
10. Zhang H, Li HC, Miao XX (2013) Feasibility, limitation and possible solutions of RNAi-based technology for insect pest control. Insect Sci 20(1):15–30. https://doi.org/10.1111/j.1744-7917.2012.01513.x
11. Clemons A, Haugen M, Le C, Mori A, Tomchaney M, Severson DW, Duman-Scheel M (2011) siRNA-mediated gene targeting in *Aedes aegypti* embryos reveals that frazzled regulates vector mosquito CNS development. PLoS One 6(1):e16730. https://doi.org/10.1371/journal.pone.0016730
12. Haugen M, Flannery E, Tomchaney M, Mori A, Behura SK, Severson DW, Duman-Scheel M (2011) Semaphorin-1a is required for *Aedes aegypti* embryonic nerve cord development. PLoS One 6(6):e21694. https://doi.org/10.1371/journal.pone.0021694
13. Nguyen C, Andrews E, Le C, Sun L, Annan Z, Clemons A, Severson DW, Duman-Scheel M (2013) Functional genetic characterization of salivary gland development in *Aedes aegypti*. EvoDevo 4(1):9. https://doi.org/10.1186/2041-9139-4-9
14. Sarro J, Andrews E, Sun L, Behura SK, Tan JC, Zeng E, Severson DW, Duman-Scheel M (2013) Requirement for commissureless2 function during dipteran insect nerve cord development. Dev Dyn 242(12):1466–1477. https://doi.org/10.1002/dvdy.24059
15. Zhang X, Mysore K, Flannery E, Michel K, Severson DW, Zhu KY, Duman-Scheel M (2015) Chitosan/interfering RNA nanoparticle mediated gene silencing in disease vector mosquito larvae. J Vis Exp (97). https://doi.org/10.3791/52523
16. Mysore K, Sun L, Tomchaney M, Sullivan G, Adams H, Piscoya AS, Severson DW, Syed Z, Duman-Scheel M (2015) siRNA-mediated silencing of doublesex during demale development of the dengue vector mosquito *Aedes aegypti*. PLoS Negl Trop Dis 9(11): e0004213. https://doi.org/10.1371/journal.pntd.0004213
17. Duman-Scheel M, Syed Z (2015) Developmental neurogenetics of sexual dimorphism in *Aedes aegypti*. Front Ecol Evol 3. https://doi.org/10.3389/fevo.2015.00061
18. Mysore K, Flannery E, Leming MT, Tomchaney M, Shi L, Sun L, O'Tousa JE, Severson DW, Duman-Scheel M (2014) Role of semaphorin-1a in the developing visual system of the disease vector mosquito *Aedes aegypti*. Dev Dyn 243(11):1457–1469. https://doi.org/10.1002/dvdy.24168
19. Whyard S, Singh AD, Wong S (2009) Ingested double-stranded RNAs can act as species-specific insecticides. Insect Biochem Mol Biol 39(11):824–832. https://doi.org/10.1016/j.ibmb.2009.09.007
20. Singh AD, Wong S, Ryan CP, Whyard S (2013) Oral delivery of double-stranded RNA in larvae of the yellow fever mosquito, *Aedes aegypti*:

implications for pest mosquito control. J Insect Sci 13(69). https://doi.org/10.1673/031.013.6901

21. Scott JG, Michel K, Bartholomay LC, Siegfried BD, Hunter WB, Smagghe G, Zhu KY, Douglas AE (2013) Towards the elements of successful insect RNAi. J Insect Physiol 59 (12):1212–1221. https://doi.org/10.1016/j.jinsphys.2013.08.014
22. Zhang X, Zhang J, Zhu KY (2010) Chitosan/double-stranded RNA nanoparticle-mediated RNA interference to silence chitin synthase genes through larval feeding in the African malaria mosquito (*Anopheles gambiae*). Insect Mol Biol 19(5):683–693. https://doi.org/10.1111/j.1365-2583.2010.01029.x
23. Mysore K, Flannery EM, Tomchaney M, Severson DW, Duman-Scheel M (2013) Disruption of *Aedes aegypti* olfactory system development through chitosan/siRNA nanoparticle targeting of semaphorin-1a. PLoS Negl Trop Dis 7 (5):e2215. https://doi.org/10.1371/journal.pntd.0002215
24. Mysore K, Andrews E, Li P, Duman-Scheel M (2014) Chitosan/siRNA nanoparticle targeting demonstrates a requirement for single-minded during larval and pupal olfactory system development of the vector mosquito *Aedes aegypti*. BMC Dev Biol 14:9. https://doi.org/10.1186/1471-213X-14-9
25. Whyard S, Erdelyan CN, Partridge AL, Singh AD, Beebe NW, Capina R (2015) Silencing the buzz: a new approach to population suppression of mosquitoes by feeding larvae double-stranded RNAs. Parasit Vectors 8:96. https://doi.org/10.1186/s13071-015-0716-6
26. Hapairai LK, Mysore K, Chen Y, Harper EI, Scheel MP, Lesnik AM, Sun L, Severson DW, Wei N, Duman-Scheel M (2017) Lure-and-kill yeast interfering RNA larvicides targeting neural genes in the human disease vector mosquito *Aedes aegypti*. Sci Rep 7(1):13223. https://doi.org/10.1038/s41598-017-13566-y
27. Mysore K, Hapairai LK, Sun L, Harper EI, Chen Y, Eggleson KK, Realey JS, Scheel ND, Severson DW, Wei N, Duman-Scheel M (2017) Yeast interfering RNA larvicides targeting neural genes induce high rates of *Anopheles* larval mortality. Malar J 16(1):461. https://doi.org/10.1186/s12936-017-2112-5
28. Mumberg D, Muller R, Funk M (1995) Yeast vectors for the controlled expression of heterologous proteins in different genetic backgrounds. Gene 156(1):119–122
29. Winzeler EA, Shoemaker DD, Astromoff A, Liang H, Anderson K, Andre B, Bangham R, Benito R, Boeke JD, Bussey H, Chu AM, Connelly C, Davis K, Dietrich F, Dow SW, El Bakkoury M, Foury F, Friend SH, Gentalen E, Giaever G, Hegemann JH, Jones T, Laub M, Liao H, Liebundguth N, Lockhart DJ, Lucau-Danila A, Lussier M, M'Rabet N, Menard P, Mittmann M, Pai C, Rebischung C, Revuelta JL, Riles L, Roberts CJ, Ross-MacDonald P, Scherens B, Snyder M, Sookhai-Mahadeo S, Storms RK, Veronneau S, Voet M, Volckaert G, Ward TR, Wysocki R, Yen GS, Yu K, Zimmermann K, Philippsen P, Johnston M, Davis RW (1999) Functional characterization of the *S. cerevisiae* genome by gene deletion and parallel analysis. Science 285 (5429):901–906
30. An C, Budd A, Kanost MR, Michel K (2011) Characterization of a regulatory unit that controls melanization and affects longevity of mosquitoes. Cellular and molecular life sciences. CMLS 68(11):1929–1939. https://doi.org/10.1007/s00018-010-0543-z
31. Clemons A, Mori A, Haugen M, Severson DW, Duman-Scheel M (2010) Culturing and egg collection of *Aedes aegypti*. Cold Spring Harb Protoc (10):pdb prot5507. doi:https://doi.org/10.1101/pdb.prot5507
32. Clonetech (2018) shRNA sequence designer tool. http://bioinfo.clontech.com/rnaidesigner/oligoDesigner.do. Accessed January 2018
33. Haugen M, Tomchaney M, Kast K, Flannery E, Clemons A, Jacowski C, Simanton Holland W, Le C, Severson D, Duman-Scheel M (2010) Whole-mount *in situ* hybridization for analysis of gene expression during *Aedes aegypti* development.. Cold Spring Harb Protoc 2010 (10): pdb.prot5509. doi: https://doi.org/10.1101/pdb.prot5509
34. Patel NH (1996) *In situ* hybridization to whole mount *Drosophila* embryos. In: Krieg PA (ed) A laboratory guide to RNA: isolation, analysis, and synthesis. Wiley-Liss, New York, pp 357–370
35. Murphy KA, Tabuloc CA, Cervantes KR, Chiu JC (2016) Ingestion of genetically modified yeast symbiont reduces fitness of an insect pest via RNA interference. Sci Rep 6:22587. https://doi.org/10.1038/srep22587

INDEX

Susan J. Brown and Michael E. Pfrender (eds.), *Insect Genomics: Methods and Protocols*, Methods in Molecular Biology, vol. 1858, https://doi.org/10.1007/978-1-4939-8775-7, © Springer Science+Business Media, LLC, part of Springer Nature 2019

O

P

Q

R

S

T

U

V

W

Z

Printed by Printforce, the Netherlands